W9-COJ-487

# The 'Deutsche Forschungsgemeinschaft' and its 'Sonderforschungsbereiche'

The Deutsche Forschungsgemeinschaft (DFG) serves science and the humanities in all fields by financing research projects. In many scientific areas, the Deutsche Forschungsgemeinschaft has taken over the task of strengthening cooperation among researchers and coordinating basic research with state research promotion. The Deutsche Forschungsgemeinschaft advises parliaments and public authorities on scientific matters, cultivates relations between science and industry and promotes cooperation between German scientists and foreign researchers. Special attention is paid to promoting young scientists.

The DFG distinguishes between funding individual projects (Individual Grants Program – Normalverfahren) and cooperative activities (Priority Programs – Schwerpunktprogramme, Research Units – Forschergruppen, Collaborative Research Centers – Sonderforschungsbereiche, Central Research Facilities – Hilfseinrichtungen der Forschung).

In Collaborative Centers groups of scientists from various disciplines combine their efforts in order to pursue a large scale investigation of a major problem. They aim at strengthening research at universities by concentrating personnel and material, coordinating research within and between universities as well as between universities and research institutions outside and by providing a long-term financial basis.

In 1991 the Deutsche Forschungsgemeinschaft funded 177 Collaborative Research Centers with more than 390 million marks. This amounts to almost a third of the DFG's total budget of DM 1.361 billion in 1991. 75 per cent of the funds are provided by the Federal Government via the Ministry of Education and Science (Bundesminister für Bildung und Wissenschaft), 25 per cent come from the sixteen states (Bundesländer).

© VCH Verlagsgesellschaft mbH, D-6940 Weinheim (Federal Republic of Germany), 1992

Distribution:
VCH, P. O. Box 10 11 61, D-6940 Weinheim (Federal Republic of Germany)
Switzerland: VCH, P. O. Box, CH-4020 Basel (Switzerland)
United Kingdom and Ireland: VCH (UK) Ltd., 8 Wellington Court,
    Cambridge CB1 1HZ (England)
USA and Canada: VCH, Suite 909, 220 East 23rd Street, New York,
    NY 10010–4606 (USA)

ISBN 3-527-27718-8 (VCH, Weinheim)
ISBN 1-56081-168-4 (VCH, New York)

ISSN 0930-4398

**DFG** Deutsche Forschungsgemeinschaft

# Molecular Biology of the Cell

Final Report of the Sonderforschungsbereich
"Molekularbiologie der Zelle" 1970–1988

Edited by
Walter Doerfler

Sonderforschungsbereich

Deutsche Forschungsgemeinschaft
Kennedyallee 40
D-5300 Bonn
Telefon: (02 28) 8 85-1
Telefax: (02 28) 8 85-22 21

Published jointly by
VCH Verlagsgesellschaft mbH, Weinheim (Federal Republic of Germany)
VCH Publishers Inc., New York, NY (USA)
Library of Congress Card No. applied for.

British Library Cataloguing-in-Publication Data:

A catalogue record for this book is available from the British Library.


Die Deutsche Bibliothek – CIP-Einheitsaufnahme

**Molecular biology of the cell :** final report of the Sonderforschungsbereich
"Molekularbiologie der Zelle" 1970 – 1988 / DFG,
Deutsche Forschungsgemeinschaft. Ed. by Walter Doerfler. –
Weinheim ; Basel (Switzerland) ; Cambridge ; New York : VCH, 1992
    (Sonderforschungsbereiche)
    ISBN 3-527-27718-8 (Weinheim …)
    ISBN 1-56081-168-4 (New York)
NE: Doerfler, Walter [Hrsg.]; Sonderforschungsbereich
    Molekularbiologie der Zelle <Köln>

Production Manager: L & J Publikations-Service GmbH, D-6940 Weinheim
Composition: Hagedornsatz GmbH, D-6806 Viernheim
Printing and Bookbinding: Wilhelm Röck, Druckerei und Großbuchbinderei,
D-7012 Weinsberg
Printed in the Federal Republic of Germany

# Contents

*Contents*

*Contents*

*Contents*

*Contents*

Contents

# 1 Introduction

# 1.1 The Sonderforschungsbereich – an Important Instrument for Research at the Universities

Worldwide, the period between 1970 and 1988 has witnessed unprecedented growth and scientific excitement in molecular biology. In a now hard-to-understand mood of nostalgia, some had predicted that the apogee in this field had been passed in the late sixties. The importance and intensity of discoveries made in the following two decades has proven such gloomy predictions wrong. During the same time span, research in molecular biology in Köln has had the good fortune of support by one of the then new programs of the Deutsche Forschungsgemeinschaft, the Sonderforschungsbereich[1] 74. In the late sixties the program of this center grant system in Köln has been named Molekularbiologie der Zelle (Molecular Biology of the Cell). This broad scope enabled many researchers who chose the thrust of their research in the field of molecular biology, to participate. The Sonderforschungsbereich 74 has provided a framework for basic research in Köln for scientists in biochemistry, plant biology, developmental biology, and genetics to name the institutional affiliations of its members. Scientists from both the University and the Max-Planck-Institut participated in the program. We all have come to realize that institute titles have gradually become obsolete and that the actual ties among scientists lie in the fascination with problems and in scientific and personal interactions directed toward their solution.

Although basic research is indigenous to the university and one of its major tasks, it has sometimes been endangered there for a number of reasons. For almost two decades, the Sonderforschungsbereich has enabled us to formulate and implement research programs in molecular and cellular biology. At the same time, it has helped scientists in Köln to provide the conditions at the university for its second equally important task, the education and formation of young scientists, starting at the diploma and doctoral levels and continuing to the young researcher who gets her or his first chance for independent research. In his much-acclaimed book "For the Love of Enzymes" Arthur Kornberg at Stanford states "If the case can be made that

---

[1] Sonderforschungsbereich = Collaborative Research Center

3

my activities in administration, writing, and teaching have made a unique contribution, then certainly a further case can be made that my discoveries in science have not. Very likely, they would have been made by others soon after". If an eminent scientist like Arthur Kornberg, who has made outstanding contributions in fundamental research and who successfully educated his students to be "obsessed" by science, ranks "administration, writing, and teaching" that high, they must be of significance for the conduct of research.


## 1.2　Sonderforschungsbereich: A Grant System or a Concept for Basic Research

It would be hypocritical not to admit that the Sonderforschungsbereich is an important source for the much needed financial support of research at university institutes. Even a very wealthy university would not be in a position to defray by itself the ever increasing costs of research in the many branches of molecular biology. However, the Sonderforschungsbereich provided much more than the material basis for our work. Molecular biologists in Köln voluntarily joined in this group effort and developed their research programs. We were very fortunate that our colleagues from the Max-Planck-Institut für Züchtungsforschung, who could afford to some extent to stay immune to the more mundane lures of Sonderforschungsbereich life, participated in the program.

The Sonderforschungsbereich 74 did not aspire to, but rather abhorred, a streamlined overall concept for research on a central topic but espoused individuality in basic research and strove to encourage a stream of different ideas, people, and projects. New project leaders were not chosen primarily for their particular research interest, but on the basis of some recent and remarkable accomplishment in science. The motto of the Sonderforschungsbereich "Molekularbiologie der Zelle" was consistent with this overall conceptual approach to the conduct of basic research in molecular biology. In at least two aspects, we were very fortunate that the activities of the Sonderforschungsbereich blended into the bustling life of one of Germany's most popular universities.

– In spite, or perhaps because of, the loose organizational structure of the Sonderforschungsbereich 74, a multitude of scientific interactions and very fruitful, long-term collaborations with joint publications have developed. Moreover, we all have appreciated the very lively exchange of ideas and techniques, the joint Friday noon seminars, and the very active program of almost daily, sometimes even more frequent, seminars by guest speakers. We all have become accustomed to taking these assets of the Sonderforschungsbereich life for granted, although they may not immediately contribute to our day-to-day research. The beneficial effects are clearly of a long-standing nature to all of us, but most importantly to our students.

– In molecular biology the research-oriented aspect of teaching has always been in the forefront of educating young scientists. In a discipline that has broken much new ground in a few decades, the encyclopedic approach to teaching and learning has fortunately not yet fogged the mind. This is not to advocate that the students in molecular biology will not have to acquire a tremendous amount of solid knowledge in biochemistry, genetics, developmental, and many other fields of biology before they can hope to identify their own research project. I venture to say that the material and ideal support of the Sonderforschungsbereich facilitated excellent diploma, doctoral, and postdoctoral work in molecular biology at the University in Köln. In that way, the Sonderforschungsbereich 74 has essentially contributed to the teaching of modern biology at this university and has thus helped to fulfill one of the major tasks of its faculty.

Before I turn to presenting an overview of the research accomplished during the past two decades, I should like to mention yet another activity of the Sonderforschungsbereich which has proved attractive far beyond the bounds of this University. The Spring Meeting in Köln has developed into a sizable international meeting, usually the first major meeting on molecular biology in Germany at the beginning of the year. Depending on the interests of its organizers, who have been members of the Sonderforschungsbereich, the invited lecturers have presented the state-of-the-art in different fields of molecular biology. These meetings were held yearly starting in 1971 (cf. Table 1.2.1 for the topics of the meetings since 1981). In launching these meetings, the emphasis has been on providing a forum for the molecular biologists in Köln to interact in a variety of ways with other scientists in the field and on stimulating also students and younger scientists in Köln to thought and discussions at an international level. Informality in organization and the last minute addition of speakers with a significant new contribution have added to the excitement and enjoyment of these meetings, at least in the biased view of someone who has helped to organize a good number of them.

**Table 1.2.1:** Spring Meetings 1981–1990.

| | |
|---|---|
| 1981 | DNA Methylation and Genome Organization<br>Organizer: Walter Doerfler |
| 1982 | Genetic Basis of Diversity<br>Organizer: Klaus Rajewsky |
| 1983 | Genetic Manipulation: Impact on Man and Society<br>Organizer: Peter Starlinger |
| 1984 | Protein-DNA Recognition<br>Organizers: Benno Müller-Hill<br>　　　　　　Albrecht Sippel |
| 1985 | Organization and Expression of the Eukaryotic Genome<br>Organizer: Walter Doerfler |
| 1986 | Molecular Biology and Evolution<br>Organizers: Konrad Beyreuther<br>　　　　　　Klaus Rajewsky<br>　　　　　　Peter Starlinger |
| 1987 | Protein-DNA Interaction<br>Organizer: Benno Müller-Hill |
| 1988 | Gene Expression and Differentiation<br>Joint Symposium of the Gene Centers Berlin, Heidelberg, Köln and München<br>Organizers: Heinrich Betz<br>　　　　　　Matthias Cramer<br>　　　　　　Walter Doerfler<br>　　　　　　Jeff Schell<br>　　　　　　Ernst-Ludwig Winnacker |
| 1989 | Neurobiology<br>Organizers: Willie Stoffel<br>　　　　　　Wolfgang Wille |
| 1990 | Recombination and Repair<br>Organizers: Walter Doerfler<br>　　　　　　Börries Kemper<br>　　　　　　Klaus Rajewsky<br>　　　　　　Peter Starlinger |

# 1.3    The Kernel of it all – the Projects

In this section, each of the major projects has been briefly summarized by the project leader. A much more detailed account of the most important results will be presented in the individual chapters of this book.

## 1.3.1    The System of *E. coli* as Problem and Tool
(Benno Müller-Hill)

Müller-Hill and his colleagues developed the *lac* repressor-operator system into a model of protein DNA recognition and repression. They isolated gram quantities of *lac* repressor and determined its protein sequence. They then demonstrated that *lac* repressor carried an N-terminal protrusion of about 50 residues, which is responsible for specific and nonspecific binding to *lac* operator. They showed that active *lac* repressor could be fused to active β-galactosidase. This led to the conclusion that two subunits of *lac* repressor are enough to recognize *lac* operator. They then demonstrated the existence of symmetric ideal *lac* operator. Using symmetric variants of ideal *lac* operator, they defined which residues of the recognition helix of *lac* repressor recognize which bases of *lac* operator. They isolated active dimeric *lac* repressor and showed that it represses only weakly. They demonstrated that wild type tetrameric *lac* repressor forms stable loops between the first and second or third *lac* operator.

They sequenced the *lac Y* permease gene and some of its mutants. They used the discovery of β-galactosidase fusions for the development of various expression vectors. They used them to clone epitopes of *P. falciparum*. They also cloned and sequenced the cDNA coding for the human amyloid precursor in an attempt to understand Alzheimer's disease.

Four independent groups worked in Müller-Hill's unit: Konrad Beyreuther, Bruno Gronenborn who analysed the expression of *Cauliflower Mosaic Virus*, Wayne Davies who sequenced the *Lambda att* site and Bernd Gutte who synthesized interesting peptides.

## 1.3.2   Membrane Lipids, Lipoproteins, and Membrane Structures
(Wilhelm Stoffel)

– Essential complex membrane lipids are the sphingolipids, derivatives of the long chain sphingosine bases as common structural elements. We unraveled the pathway of its biosynthesis and degradation. The results have become general text book knowledge. Interestingly, the degradation products palmitaldehyde, hexadeca-2-enal and phosphoryl-ethanolamine are metabolic links between sphingosine catabolism and phospholipid biosynthesis, particularly plasmalogens.

– The development of synthetic $^{13}$C-labeled fatty acids, cholesterol, [$^{13}$CH$_3$-N-choline] phospholipids for $^{13}$C-NMR spectroscopy together with chemical crosslinking studies led to the presently accepted model of the serum high-density lipoprotein particle.

– Cell biology studies *in vitro* unravelled the synthesis, translocation, secretion, and processing of apo AI and AII. Their cDNA clones and human genes have been isolated, and the regulation of the expression of these two genes are presently being studied.

– The two predominant proteins of rat, bovine, and human myelin membranes of CNS have been analyzed, the structure of the integral membrane proteolipid protein (PLP) elucidated, its cDNA and that of myelin basic protein (MBP) cloned, the human gene organisation of both determined and the X-chromosomal locus determined. On this basis animal and human dysmyelinoses have been analyzed by recombinant DNA techniques: the jimpy mouse myelin deficient rat and Pelizaeus-Merzbacher (PM) disease turned out to be the result of point mutations leading to a missense PLP protein (ji) or disturbed PLP conformations (md rat and PM-disease) with the death of oligodendrocytes. These studies now allow the analysis of one of the most important differentiation processes in CNS, that of myelinogenesis by oligodendrocytes.

## 1.3.3  Resolution of Recombination Intermediates
(Börries Kemper)

Recent, popular models of homologous genetic recombination propose the Holliday-structure as a key intermediate. The existence of the structure was originally predicted by Robin Holliday as a part of a model based on genetic data from crosses in fungi. The crossover-structure was described as two homologous, double-stranded DNA molecules which, during synapsis, have mutually exchanged homologous strands. Its enzymatic resolution is an intrinsic part of the model because of its vital importance for the survival of recombining cells.

Since its original proposal in 1964, most of the basic mechanistic predictions of Holliday's model were proven to be real in enzymatic terms *in vitro*. However, the key question of how a Holliday-structure could enzymatically be resolved, remained unclear until recently, when endonuclease VII was isolated in our laboratory. It is the first enzyme reported to resolve Holliday-structures *in vitro*. This was taken in support of the existence of specific Holliday-structure resolving enzymes.

Endonuclease VII is the product of gene 49 of bacteriophage T4, and its activity is required *in vivo* to remove branches from newly synthesized DNA. In its absence, the DNA cannot be packaged to completeness, and empty heads accumulate in the infected cell. Gene 49 was recently cloned, overexpressed in *E. coli*, and its product purified to homogeneity. The active protein is a dimer with monomers of molecular weight of 18kDa. This agrees well with the predicted molecular weight of 18.12kDa deduced from the sequence of the gene. The purified protein converts branched DNAs in Holliday-structures into linear molecules with high precision, producing genetically meaningful products. The specificity of endonuclease VII, however, is not restricted to Holliday-structures, and a broad range of other structural deformations of DNA duplexes were also shown to trigger the enzyme. This includes heteroduplex loops, Y-junctions, mismatches, and single-stranded overhangs.

A search for comparable enzyme(s) in extracts from the yeast *S. cerevisiae* led us to the discovery of an eukaryotic protein with nearly the same characteristics. Recent reports from other laboratories describe Holliday-structure resolving enzymes in cells from higher organisms. The findings confirm our belief that Holliday-structure resolving enzymes are ubiquitous enzymes, indeed, and not just a specialty of bacteriophage. Enzymes with the described specificities were named X-solvases to distinguish them from transposon related resolvases, with endonuclease VII as the prototype.

## 1.3.4 Transposable Elements in Bacteria and in Plants

(Peter Starlinger, Heinz Saedler)

Since the end of the sixties and particularly during the seventies when the Sonderforschungsbereich was founded by the Deutsche Forschungsgemeinschaft, Heinz Saedler and Peter Starlinger, together with their collaborators, elucidated the structure of unusually polar mutations in bacteria. They found that these were insertions of DNA segments of limited length. The sequences that were independently discovered by J. Shapiro are today known as IS-elements and belong to the large class of transposable elements. Genetically, they resemble the transposable controlling elements in maize that were discovered by Barbara McClintock. IS-elements belong to a large class, of which particularly those carrying an antibiotic resistance have become interesting. In Köln, IS1, IS2, and IS4 were studied.

Since the end of the seventies, the work was extended to plant transposable elements. At the Max-Planck-Institut für Züchtungsforschung, Heinz Saedler and his collaborators study the maize element *En/Spm*. At the Institut für Genetik, the maize transposable element *Ac* is studied.

## 1.3.5 *Agrobacterium*-Mediated Gene Transfer to Plants

(Jeff Schell)

The interaction between *Agrobacterium tumefaciens* or *Agrobacterium rhizogenes* and susceptible plant cells presents a number of fundamentally interesting biological questions that stem from the ability of *Agrobacterium* to transfer a particular DNA segment, the T-DNA, from the Ti or Ri plasmid of the bacterium to the plant nuclear genome. The fact that *Agrobacterium* genetically transforms plant cells has led to intensive investigations of how DNA transfer occurs and the development of modifications of the system which allow the genetic engineering of plant cells with any DNA of interest. The equally intriguing question of how the natural *Agrobacterium* T-DNA element specifies neoplasia, termed crown gall, has been studied continuously since the beginning of the century. Using *Agrobacterium* strains devoid of tumor-inducing functions and carrying a selectable marker gene flanked by the 25 bp integration sequences, it has been possible to obtain a large number of transgenic plants, e. g. in tobacco, petunia, tomato, potato amongst others. The integrated DNA is stable and inherited according to simple Mendelian segregation patterns. Gene transfer methods have been used successfully to study the molecular mechanisms underlying regulated gene expression in plants. It has been found that chimeric genes combining an alien coding sequence with regulatory sequences derived from defined plant genes can function in predictable fashion in transgenic plants. These observations open the way for a systematic use of gene transfer techniques in plant breeding.

## 1.3.6　Resistance Mechanisms in Plants
(Klaus Hahlbrock)

Research in this group is focused mainly on the molecular mechanisms of disease resistance and other stress responses in plants. The long-term goal is to understand the mode of action of genes conferring resistance against fungal and other pathogens. Detailed fundamental knowledge in this area is expected to provide the basis for improved breeding strategies. These strategies include the use of new gene transfer techniques, as well as improved methods of crop protection in accordance with ecological needs. We also investigate the mechanism of UV protection in plants.

Three typical plant-pathogen interactions are being studied in detail: the non-host resistance response of parsley to the soybean pathogen *Phytophtora megasperma*; the R-gene-dependent host resistance or susceptibility of the potato to one of its most destructive fungal pathogens, *Phytophtora infestans* (causal agent of late blight); and the interaction of barley with *Rhynchosporium secalis* which might involve the action of a host-selective fungal toxin.

These investigations involve the use of plant cell cultures and fungal or other elicitors (signal molecules triggering the plant's defense response); histochemical and immunocytological localization of defense-related compounds in infected plant tissue; the structural identification of such compounds and the elucidation of their biosynthetic pathways: attempts to identify effector-binding molecules using monoclonal antibodies; *in situ* hybridization of induced mRNAs; cloning and structural characterization of genes activated in infected cells; gene transfer and the identification of factors regulating transcriptional activity.

### 1.3.7  Neurogenesis in *Drosophila*
(José Campos-Ortega)

We approach the problem of the origin of cell diversity by studying the mechanisms of commitment of the neural progenitor cells in *Drosophila melanogaster*. 2000 cells in the neurogenic region of the ectoderm adopt one of two different fates: 500 cells develop as neuroblasts and the remaining 1500 cells develop as epidermal progenitor cells, or epidermoblasts. Neighboring cells must decide between two alternative developmental fates: the neural and the epidermal. We are trying to understand how this process is regulated. The decision of *Drosophila* neuroectodermal cells for one of the two lineages depends on interactions with other cells in the neighborhood. During the last several years, we succeeded in identifying the genes whose products mediate these interactions (required for the cells to take on one fate or the other). We have studied the functional interrelationships between the genes, cloned the DNA of some of them, and established the sequence of the putative gene products. The encoded proteins form a complex functional network. Its main elements are two transmembrane proteins, encoded by the genes *Notch* and *Delta*, which pass regulatory signals between the neuroectodermal cells, and two groups of putative regulatory, DNA-binding proteins, encoded by the genes of the *achaete-Scute* complex (and others) and of the *Enhancer of split* complex that controls the neural and epidermal pathways of development, respectively, in response to these signals. Several cellular, genetic, and molecular aspects of this process are discussed.

### 1.3.8  Differentiation, Pattern Formation, and Sexual Induction in *Volvox carteri*
(Lothar Jaenicke)

*Volvox carteri* is a simple and, at the same time, a sufficiently evolved eukaryont for investigating differentiation and morphogenesis at a biochemical level which we did for the past score years. Volvox is a dioecious colonial green flagellate composed of only two types of cells, germ-line gonidia and somatic body cells, and it is haploid most of the time. By an outside signal, excreted by the mature males, sexual progeny, *viz* immotile eggs and flagellated spermatozoids, respectively, instead of vegetative gonidia, is produced. Released sperm cells penetrate the somatic hull of the female spheroids and fuse with the eggs forming the diploid zygote, which, after meiosis, gives rise to a new vegetative generation. The inducing glycoprotein was isolated, its protein chain characterized via its cDNA, and the N̄- and O-glucan moieties analyzed. The N-glycosylating oligosaccharides are essential for biological activity. The external signal is transposed into an internal message that is received by the genome switching its information output to change during development from the vegetative cleavage pattern into the sexual one, typified by postponing the unequal division, setting aside sexual cells instead of daughter gonidia. The internal signal chain is accompanied by intermittent changes in the concentration of cyclic adenylate. Members of a cAMP regulatory cascade have been characterized, yet without final proof of direct connection to the morphogenetic event. Also the enzymes necessary for dissolving somatic sheath and female cell wall have been characterized in the female and in the male progeny. Cell walls of the volvocoid flagellates are composed of glycoproteins, and the lytic enzymes are individual proteases of high specificity for the phase-typic cell wall. Studies with *Chlamydomonas*, the primordial organism of the class, show an intricate regulatory network of phase-specific changes in cell wall glycosylation, autolysin production, and autolysin specificity. The molecular events are governed by the genome depending on environmental factors. Despite the present lack of sufficient genetic knowledge the green flagellates prove to be excellent and challenging objects for the cell biologically minded biochemist.

## 1.3.9 Selection of the Receptor Repertoire in the Immune System

(Klaus Rajewsky)

On the basis of experiments using hapten-carrier conjugates, we developed, in parallel with Avrion Mitchison in London, the antigen-bridge model, the first to describe a mechanism of the collaboration of T helper and B lymphocytes. Subsequent experiments showed that helper cells, as well as B cells, are targets of immunological tolerance and that helper cells exhibit exquisite antigen binding specificity and can also interact with B cells in the absence of antigen through direct receptor-receptor interactions. This result was one example of idiotypic interactions in the immune system, a subject which we investigated in collaboration with Klaus Eichmann's group. Our experiments demonstrated that network regulation as postulated by Jerne can operate at various levels in experimental model systems. Small doses of anti-idiotypic antibodies specifically regulated both T and B cell activities and dramatically affected the expression of the corresponding (complementary) idiotypes, either by suppression or by enhancement. This led to a new approach to vaccination based on anti-idiotypic antibodies. Using methods of molecular biology we determined the structure of complementary antibodies in the idiotypic network and discovered that these antibodies often carried somatically mutated variable regions.

This led us into a molecular analysis of B cell differentiation and the selection of the antibody repertoire. We found primary response antibodies to be encoded by germline genes, whereas in the secondary response somatic antibody mutants are expressed. We showed that somatic mutation is directly responsible for the affinity maturation of antibodies in the course of immunization and that it is restricted to a particular pathway of B cell differentiation, namely that of the generation of memory cells. Another process affecting antibody structure in the course of the immune response is that of the antibody class switch. The work of Andreas Radbruch and other members of the group demonstrated that B cells can be programmed, presumably by exogenous signals, to switch to a particular antibody class ("directed class switching"). In other differentiation pathways, B cells express a stable repertoire of antibodies. In the population of so-called Ly1 B cells, a set of self-propagating cells producing "natural" antibodies, we demonstrated the expression of a restricted repertoire of germline antibodies of the IgM class.

Much of the above work was only possible through a combination of methods of molecular and cell biology. The latter approach included cell separation techniques and, in particular flow cytometry and fluorescence activated cell sorting (FACS), which became a main strength of the laboratory through the activity of Andreas Radbruch and co-workers. FACS allowed us the isolation of rare somatic cell mutants, which gave clues, e. g., to mechanisms of somatic hypermutation and of class switching of antibodies and were also of practical use. This method was also instrumental in the discovery of a new protein required for the expression of antibodies on the cell surface and was probably involved in signalling through the antibody receptor by Michael Reth's work and that of our group.

Four independent junior groups worked in association with Klaus Rajewsky's unit: Klaus Eichmann, who discovered idiotype-specific suppressor T cells; Thereza Imanishi-Kari, who used idiotypic markers to study antibody V region genes and idiotypic regulation; Günther Hämmerling, who produced the first monoclonal antibodies to murine major histocompatibility (MHC) antigens and Hidde Plough, who worked on the structure, glycosylation and expression of MHC molecules.

## 1.3.10   Viruses as Troyan Horses in Molecular Biology
### (Walter Doerfler)

For several decades adenoviruses have been an excellent tool in the elucidation of elementary mechanisms in mammalian cells and have also held a prominent place in work on the mechanism of DNA viral oncogenesis.

– The integration of the viral DNA and its persistence in the genome of the transformed or tumor cell is one of the characteristics of adenoviral oncogenesis. The mechanism of the insertion of viral DNA into the host genome has been investigated as a model for the uptake and insertion of foreign DNA into the genomes of mammalian cells. Our work on adenovirus DNA integration has been among the first demonstrations of this event.
– In a study on the differential expression of the integrated adenoviral genes in transformed and tumor cells, the role of sequence-specific promoter methylation in the long-term silencing of eukaryotic genes has been discovered.
– The major late promoter of adenovirus type 12 provides the opportunity to study factors influencing the species-specific expression of this promoter.
– The regulation of expression of the E1 region of adenovirus DNA has been investigated (H. Esche).
– Adenovirus DNA has been developed into one of the best understood mammalian replicons. The biochemistry of its replication has been studied in cell free systems (E.-L. Winnacker).
– Sindbis virus (D. T. Brown) and baculoviruses (S. T. Tjia and W. Doerfler) have also been investigated, and details of viral gene expression have been elucidated in insect cells. The baculovirus *Autographa californica* has been widely used as an efficient eukaryotic expression vector system.

## 1.3.11   Alzheimer's Disease
### (Konrad Beyreuther)

In Table 1.3.1 all projects and project leaders supported by the Sonderforschungsbereich 74 during the period of 1970 to 1988 have been listed.

**Table 1.3.1:** Sonderforschungsbereich 74: Its members and their projects*.

| Name and Project | Period Associated with the Sonderforschungsbereich 74 |
|---|---|
| *Berek, Claudia*<br>Studies on the genetic basis of antibody diversity | 1986–1988 |
| *Besemer, Jürgen*<br>Biochemical characterization of mutations in the insertion sequences IS1 and IS2 | 1976–1980 |
| *Beyreuther, Konrad*<br>Molecular biology of proteins (Protein structure, protein synthesis and Alzheimer's disease) | 1969–1987 |
| *Brown, Dennis T.*<br>The development of arthropod-borne viruses in vertebrate hosts and in the invertebrate vector | 1973–1978 |
| *Campos-Ortega, José A.*<br>Studies on the genetic and molecular basis of early neurogenesis of *Drosophila* | 1983–1988 |
| *Christalla, Volker*<br>Structure and function of factors involved in DNA replication | 1977–1980 |
| *Cramer, Matthias*<br>Photo-activateable reagents in the analysis of immunological interactions | 1982–1988 |
| *Davies, R. Wayne*<br>The DNA sequence of the lambda attachment site | 1971–1977 |
| *Diers, Lothar*<br>Mitochondria and plastids | 1969–1971 |
| *Doerfler, Walter*<br>On the mechanisms of adenoviral (foreign) DNA integration and of the regulation in viral gene expression | 1972–1988 |
| *Döring, Hans-Peter*<br>Genetic and biochemical characterization of Ds and Ac elements | 1985–1988 |
| *Ehring, Ruth*<br>Regulation of the *gal* operon: Site-specific mutagenesis | 1971–1988 |
| *Eichmannn, Klaus*<br>Genetics and differentiation of antibody-producing lymphocyte clones in mice | 1971–1976 |
| *Emmerich, Hans*<br>On the mechanism of activation of ecdyson in dipteric giant chromosomes | 1970–1973 |
| *Esche, Helmut*<br>Structure and function of adenovirus-specific tumor antigens | 1980–1985 |

**Table 1.3.1:** (continued)

| Name and Project | Period Associated with the Sonderforschungsbereich 74 |
| --- | --- |
| *Fritz, Hans-Joachim*<br>Chemical synthesis of 2'-deoxyoligonucleotides<br>and their application in site-directed mutagenesis | 1978–1984 |
| *Gronenborn, Bruno*<br>Phage M13 as a tool for genetic manipulations on<br>caulimo viruses and higher plants | 1978–1982 |
| *Gutte, Bernd*<br>Structure and function of proteins | 1971–1977 |
| *Halbrock, Klaus*<br>Molecular mechanisms in disease resistance and stress<br>response in plants | 1985–1988 |
| *Hämmerling, Günther*<br>The structure of the T cell receptor and idiotypic<br>manipulations of the response to surface antigens | 1976–1979 |
| *Hermann, Klaus-Otto; Lengeler, Joseph;*<br>*Unsöld, Hans-Jürgen*<br>Transport (processes) of galactose through the<br>bacterial membrane | 1969–1972 |
| *Imanishi-Kari, Thereza*<br>Variable region gene expression and B cell diversity | 1978–1981 |
| *Jaenicke, Lothar*<br>The biochemistry of differentiation and morphogenesis<br>in *Volvox carteri* | 1969–1988 |
| *Kemper, Börries*<br>The biochemistry of genetic recombination:<br>Endonuclease VII | 1973–1988 |
| *Kneser, Hubert*<br>Interactions of RNA polymerase with different<br>promoters | 1969–1980 |
| *Kölsch, Eckehart*<br>Macrophages and the induction of the immunological<br>response | 1969–1971 |
| *Krawinkel, Ulrich*<br>The rearrangements of immunoglobulin genes in B<br>lymphocyte genes in B lymphocytes | 1981–1988 |
| *Müller-Hill, Benno*<br>Structural and functional analysis of selected proteins<br>by the manipulation of their DNA | 1969–1988 |

**Table 1.3.1:** (continued)

| Name and Project | Period Associated with the Sonderforschungsbereich 74 |
|---|---|
| *Overath, Peter*<br>The genetics of lipid and membrane metabolism in bacteria | 1969–1973 |
| *Peters, J. Hinrich*<br>Interactions between macrophages and lymphocytes | 1975–1980 |
| *Pfeifer, Dietrich*<br>Illegitimate recombination in the galactose operon of *E. coli* | 1969–1980 |
| *Ploegh, Hidde*<br>Molecular biology of the major histocompatibility complex | 1981–1984 |
| *Rajewsky, Klaus*<br>Selection of the receptor repertoir in the immune system | 1969–1988 |
| *Ruppel, Hans-Georg*<br>Studies on the formation of thylakoid membranes | 1970–1977 |
| *Saedler, Heinz*<br>Biology of the DNA sequences IS1 and IS2 | 1971–1988 |
| *Saumweber, Harald*<br>The function of specific proteins during the regulation of gene expression in *Drosophila* | 1985–1988 |
| *Schell, Jeff*<br>The mechanism of integration of the tumor inducing DNA from *A. tumefaciens* in transformed plant cells | 1978–1988 |
| *Sippel, Albrecht*<br>Regulation of gene expression by steroid hormones | 1979–1983 |
| *Starlinger, Peter*<br>Molecular biological studies on maize genes | 1969–1988 |
| *Stoffel, Wilhelm*<br>Physical and chemical studies on artificial and natural membranes and on serum lipoproteins | 1969–1988 |
| *Vielmetter, Walter*<br>Structure and function of specific DNA domains on the *E. coli chromosome* | 1969–1980 |
| *Willecke, Klaus*<br>Gene transfer and regulation of gene expression in mammalian cells | 1971–1978 |
| *Winnacker, Ernst-Ludwig*<br>Adenovirus DNA replication | 1973–1977 |

*The abbreviated titles listed are those of the most recent projects.

# 1.4 Independent Investigators – a Novel Element in the Sonderforschungsbereich 74

In the late sixties, the Max-Planck-Society pioneered a system in which young researchers were provided with the opportunity for independent research during a period of five years. This system was adopted in a similar fashion by the Sonderforschungsbereich 74, practically from the inception of its work. This element of regularly changing investigators of quite different scientific backgrounds and with new ideas, contacts, and interests has added additional ferment to the activities of the Sonderforschungsbereich. Over the past twenty years, we have been extremely fortunate in that these very talented colleagues worked on a wide scope of different subjects, which sometimes complemented in an ideal way the activities of some of the permanent groups. Table 1.3.2 reviews the names and present responsibilities of all those former members of the Sonderforschungsbereich who have moved on to positions at other universities or research institutes.

**Table 1.3.2:** Independent investigators of the Sonderforschungsbereich 74 – present positions.

| Name | From – to | Present Responsibility |
|------|-----------|------------------------|
| *E. Kölsch* | 1969–1971 | Professor for Immunology, Universität Münster |
| *P. Overath* | 1969–1973 | Director at the Max-Planck-Institut für Biologie, Tübingen |
| *H. Emmerich* | 1970–1973 | Professor for Zoology, Technische Universität Darmstadt |
| *H. G. Ruppel* | 1970–1977 | Professor for Botany, Universität Bielefeld |
| *H. Saedler* | 1971–1975 | Director at the Max-Planck-Institut für Züchtungsforschung, Köln |
| *K. Eichmann* | 1971–1976 | Director at the Max-Planck-Institut für Immunbiologie, Freiburg |
| *B. Gutte* | 1971–1977 | Professor, Universität Zürich |
| *R. W. Davies* | 1971–1977 | Senior Lecturer, University of Manchester, Department of Biochemistry |

**Table 1.3.2:** (continued)

| Name and Project | Period Associated with the Sonderforschungsbereich 74 | |
|---|---|---|
| *L. Diers* | 1969–1971 | Professor for Biology and its teaching, Universität zu Köln |
| *K. Willecke* | 1971–1978 | Professor for Genetics, Universität Bonn |
| *D. T. Brown* | 1973–1978 | Professor and Director, the Cell Research Institute, University of Texas at Austin |
| *E. L. Winnacker* | 1973–1977 | Professor for Biochemistry, Universität München (LMU), Director at the Genzentrum München |
| *H. J. Peters* | 1975–1980 | Professor for Hygiene and Human Genetics, Universität Göttingen |
| *G. Hämmerling* | 1976–1979 | Head of department and Professor at the Deutsches Krebsforschungszentrum Heidelberg |
| *V. Christalla* | 1977–1980 | Professor, Fachhochschule Braunschweig-Wolfenbüttel |
| *T. Imanishi-Kari* | 1978–1981 | Associate Professor, Tufts University Boston |
| *B. Gronenborn* | 1978–1982 | Research Director, Institut des Sciences Végétales, Gif-sur-Yvette |
| *A. E. Sippel* | 1979–1983 | Professor, Institut für Biologie III, Universität Freiburg |
| *H. Esche* | 1980–1985 | Professor for Molecular Biology, Universität Essen |
| *H. J. Fritz* | 1980–1984 | Professor for Genetics, Universität Göttingen |
| *H. L. Ploegh* | 1981–1984 | Head of department, Antonie van Leeuwenhoekhuis, Het Nederlands Krankerinstituut, Amsterdam |
| *K. Beyreuther* | 1969–1987 | Professor for Molecular Biology, Universität Heidelberg (ZMBH) |
| *U. Krawinkel* | 1981–1988 | Head of a research unit, Universität Freiburg |

# 1.5    The Next Grant Application is Due

The adage "Aus der Not eine Tugend machen" applies to one of the formalistic, but nevertheless essential, aspects of daily life in the Sonderforschungsbereich, namely writing and defending the next grant application. After almost 25 years of practically ceaseless writing of grant applications, it is difficult to suppress the clandestine question, "Do they want you to do it again?" On the other hand, an important case can be made for the notion that a formal application assists in the sorting out of ideas, forces to distinguish the essential from the peripheral, and allows a critical review of the work during the preceding years. Perhaps, it is also true that genuinely new ideas are usually developed under different circumstances. A grant application could, but does not have to, be anathema to informal thoughts and novel insights.

Perhaps, the Deutsche Forschungsgemeinschaft should seriously consider expanding the usual three year cycle to one of at least five years. Such a regime would lessen the burden on applicants and reviewers alike and hardly reduce productivity and critical judgment. However, we have tried to make the best of a necessity, and the need for regular peer review has, in a final analysis, helped the Sonderforschungsbereich considerably. We are most grateful to the reviewers who have visited us on site every three years, frequently made useful suggestions and kept critical eyes on our endeavors, often in sessions that lasted many hours past midnight. Thanks are also due to our colleagues of the Deutsche Forschungsgemeinschaft who bore the brunt of the painstaking accounting and of keeping the rules in the system.

# 1.6    The Future of the
#         Sonderforschungsbereich 74

One of these rules in the Deutsche Forschungsgemeinschaft stipulates that a Sonderforschungsbereich has to be put to rest after a certain number of years. So we folded our portfolio and started to write this report. At this somber moment, very mysteriously, the Sonderforschungsbereich commenced to demonstrate its very biological potential. In the middle of a lot of counterproductive bureaucratic hustle, the Sonderforschungsbereich divided mitotically and two new Sonderforschungsbereiche evolved: Sonderforschungsbereich 243 (Molecular Analysis of the Development of Cellular Systems) and Sonderforschungsbereich 274 (The Modular Structure of the Genetic Material). At least in this cytological sense then the Sonderforschungsbereich 74 has reached its destination and has even become a success. I shall forego the conventional temptation to enumerate the tally of scientific publications, of diplomas, doctoral theses, prizes, and honors that this Sonderforschungsbereich has helped to generate. It has not done badly on this account either. Even "Habilitationen", these uniquely Teutonic exercises without scientific significance but of substantial academic sequelae, will remain unmentioned. All of us have helped to train a large number of doctoral students, many of whom today hold professorial or equivalent positions and actively conduct their own research in various branches of molecular biology. One consequence of this multiplication effect is the ever increasing number of grant applications in biology that the Deutsche Forschungsgemeinschaft will have to cope with. I consider that phenomenon a very healthy sign of research activities in Germany. This challenge calls for an even broader financial basis and wider support for high quality basic research at the universities in the now reunited Germany. At the same time, it is realized that growth cannot be unlimited and quality control, also at the level of doctoral theses, has to remain tight.

# 1.7 The Direction of Research in Molecular Biology in the Nineties

As mentioned in the first paragraph of this preface, predictions about the direction of research are too often wrong and eventually compromise the prophet-to-be. As one might be able to foresee now, molecular biology is still holding tremendous potential for the discovery of fundamental processes. Success will not come easily and research will be more complicated and costly than in the early days. The systems under investigation only now reveal their tremendous complexities. Already for many years, molecular biology has been directed towards medicine, to plant biology, to developmental biology, and to other fields. Applied research will assume an ever increasing role.

For very complicated and hard-to-understand reasons molecular biology in Germany has had and perhaps will have to cope with a minority, but very vociferous, opposition in the public which is often poorly informed but all the more actively attempts to arouse public sentiment against this and other fields of fundamental research, in many cases for obvious, short-term political gains. Unfortunately, even responsible politicians have not always been immune to these shrill calls from the emotional abyss. It would be naive to ignore these potentially very dangerous currents, although one must not overreact. Nevertheless, many of us have devoted considerable time to lectures for the public, a much needed and useful endeavor, and to committee work, an ambivalent, if unavoidable exercise. "Nubeculum est, transibit?" A rational, cautious and courageous approach is called for that will enable us to continue to do essential basic research in a sense that we think will be in the best interest of the public.

*Acknowledgements*

We are greatly indebted to all branches of the Deutsche Forschungsgemeinschaft for their long-standing, effective, and unbureaucratic support. We also thank the "Universität zu Köln" and the "Ministerium für Wissenschaft und Forschung des Landes Nordrhein-Westfalen" for providing the very substantial framework for basic research at the University institutes in Köln. The "Stiftung Volkswagenwerk" has enabled the Institute of Genetics to expand laboratory space at an extremely critical moment of growth. The "Bundesministerium für Bildung und Wissenschaft" (via the Deutsche Forschungsgemeinschaft) and "Bundesministerium für Forschung und Technologie" (via the Max-Planck-Gesellschaft and through several of the Ministry programs, mainly through the creation of the "Genzentrum Köln") have provided essential support for many branches of molecular biology in Köln. The following have helped on an individual but very remarkable and extremely helpful basis: Bayer AG (Bayer professorship), Fonds der Chemischen Industrie, Fritz-Thyssen-Stiftung (Graduiertenkolleg), Stiftung Fazit, Hoechst AG, and others.

My colleagues in the Sonderforschungsbereich have been most supportive and have always shown this science-oriented "esprit de corps" which rendered daily collaboration pleasant and constructive. The speaker of a Sonderforschungsbereich has a tough job because he has to deal with many high-powered "prime donne"; it is challenging because most activities are related to research, although usually through the filter of financial responsibilities. The job, to be absolutely frank, would have been impossible without the collaboration, advice, the competence, the uncompromisingly critical attitude, and the daily hard work of Anne Weber. In many ways she has made essential contributions to the efficiency of the Sonderforschungsbereich, and all that in a very pleasant atmosphere.

I also thank the present and former members of the Sonderforschungsbereich 74 for submitting their manuscripts on time – or not more than six months past the assigned deadline. The final report is the hardest of them all because it has to dwell on excitements of the "distant past".

Köln, May 1991                                                 Walter Doerfler

# 2 Molecular Mechanisms

# 2.1 The *lac* System of *E. coli* as Problem and Tool

Benno Müller-Hill[*]

## 2.1.1 Introduction: The *lac* System as a Paradigm of Molecular Biology

The lactose system of *E. coli* became a paradigm of molecular biology the moment Jacques Monod touched it. During the war, he worked in the Sorbonne in Paris devoting half of his time to science and half to the resistance. Yet he found the time to discover two fundamental aspects of the system: diauxy [1] and *lac* constitutive mutants [1]. Diauxisy is easy to explain. When *E. coli* is grown on a mixture of glucose and lactose, it first uses all the glucose. Growth stops for a few hours and then begins again, using the lactose. Monod showed that *E. coli* grown on glucose contains very little β-galactosidase, the enzyme which breaks down lactose. But when grown on lactose its content of β-galactosidase increases dramatically. He called this phenomen "induction". When one transfers the bacteria back to glucose, the level of β-galactosidase is diluted to the initial low value. Monod called the phenomenon of stepwise growth diauxy. *lac* constitutive mutants, which produce high amounts of β-galactosidase all the time, escape diauxy; they can use glucose and lactose at the same time.

How is the increase in induction, i. e. increase in synthesis of β-galactosidase, brought about? Monod and his collaborators showed that it occurs by *de novo* synthesis and not by folding of already existing polypeptide chains [1]. In 1961, François Jacob and Jacques Monod deciphered the control system [1]. It consists of a *repressor*, which binds in the absence of an *inducer* (lactose) to an *operator* located close to the *promoter*, the site where transcription of the mRNA coding for the β-galactosidase begins. The repressor thus inhibits the start of transcription. In the presence of inducer, the *lac* repressor does not bind to the operator and the synthesis of β-galactosidase can begin.

---

[*] Institut für Genetik der Universität zu Köln

I entered the field as a graduate student of Kurt Wallenfels (Universität Freiburg, FRG) working on the structure of β-galactosidase. I continued work on the *lac* system as a postdoctoral fellow, first with Howard Rickenberg (Indiana University, Bloomington Ind., USA) and then with Walter Gilbert (Harvard University, Cambridge, Mass., USA). Together with Gilbert, I isolated the *lac* repressor [2] and constructed a mutant of the *lac I* (repressor) gene which overproduced *lac* repressor 300 fold [3]. So I knew exactly what I wanted to do when I moved to Köln in 1968, where the Sonderforschungsbereich 74 was founded two years later. I wanted to understand the molecular structure of the various parts of this system.

Of course, I was not alone working on the *lac* system. The space limit and the time span to be covered make it impossible to cite here even some of the most important papers of other groups. I may just mention that the collected papers of Monod have been printed as book [1] and that one book has been devoted *in toto* [4] and another one in a large part [5] to the *lac* system.

## 2.1.2 The *lac* Repressor *lac* Operator System

### 2.1.2.1 The Primary Structure of *lac* and *gal* Repressor

When the Sonderforschungsbereich 74 was founded in Köln in 1970, I had set for myself two goals: first, to elucidate the primary structure of *lac* repressor and second, to define by genetic analysis of mutants of the *lac I* gene the parts of *lac* repressor which were responsible for operator binding, inducer binding, and dimer and tetramer formation. I had created the technical possibility for these ventures in Harvard. The over-producer I had constructed there produces 300 times more *lac* repressor than wild type does. In 1969 I recruited Konrad Beyreuther as a postdoctoral fellow and three graduate students (Klaus Adler, Norbert Geisler and Alex Klemm) to do the protein sequencing. Starting from several kilograms of cells, they isolated about ten grams of *lac* repressor for protein sequence analysis, reported by Beyreuther in 1973 [6]. In some sense, the structure looked disappointing. The *lac* repressor looked like a normal protein. Inspection of the amino acid sequence did not suggest any peculiar structure which might bind the *lac* operator. We used a genetic approach to search for such a binding domain (see section 2.1.2.5). Only the *lac* repressor and the *lambda cro* protein have been sequenced as proteins. All other repressors whose sequences are known have been sequenced as DNA. A graduate student of mine, Brigitte von Wilcken-Bergmann, later sequenced the gene coding for *gal* repressor [7]. To our pleasant surprise, its sequence was similar to the sequence of the gene coding for *lac* repressor. Today we know five other repressors or activators of *E. coli*, which are similar in sequence to *lac* and *gal* repressor.

28

In addition, I noted that three sugar-binding, periplasmic proteins are structurally similar to the inducer binding regions of all these repressors [8]. As I will show later, the knowledge of such structurally similar systems was of great help to decipher the functioning of residues of *lac* repressor (see section 2.1.2.5).

## 2.1.2.2   The Quarternary Structure of *lac* Repressor

Both *lac* repressor and β-galactosidase are tetrameric proteins, which consist, even at high dilution, of four identical subunits. So it came as a surprise when I isolated fusions between active *lac* repressor and active β-galactosidase. My graduate student Jürgen Kania showed that the fusion protein was itself a tetramer [9]. One such repressor-β-galactosidase fusion was later sequenced by Irvin Zabin and his collaborators. They showed that the fusion was lacking the last six residues of *lac* repressor and the first 17 residues of β-galactosidase [10]. This and all other such fusion proteins were themselves tetrameric. How were the subunits aggregated in the fusion protein? Denis Brown and Jürgen Kania [11] could show that they were aggregated via the β-galactosidase part. They concluded that two subunits of *lac* repressor are sufficient for operator recognition. Kania further strengthened this argument by analysing mixed repressor-β-galactosidase tetramers, which contained various amounts of β-galactosidase subunits. Again two repressor-β-galactosidase subunits lead to appreciable, although not full, operator binding [12].

   *lac* repressor remains a tetramer even at very low concentrations. By contrast, *gal* repressor is a dimer even at very high concentrations. My students Stefan Oehler and Siegfried Alberti recently found what makes these dimers stick to each other to form tetramers: A frameshift mutation in codon 330 (of 360) the *lac I* gene reduces *lac* repressor to a dimer [13]. Oehler and Alberti recently defined the residues, which are responsible for aggregation. It is the sequence 342 *L A D S L M Q L A R Q V S R L* 356. The replacement of any one of the leucines or the valine by alanine drastically reduces aggregation. The other residues may be replaced by alanine without showing such a large effect. The structure fulfills the sterical demand of the formation of a coiled coil and thus resembles a leucine zipper. And indeed, when Oehler put the tail of *lac* repressor onto *gal* repressor he made it form a tetramer [13].

## 2.1.2.3   The Mechanism of Repression of the *lac* System

Genetics, and later DNA sequence analysis, had shown *lac* operator to be situated just downstream of *lac* promoter. This suggested a very simple mode of action of *lac* repressor: steric hindrance of RNA polymerase. This picture was not in essence changed after two other operator-like sequences ("pseudo-operators") were found to lie 92 bp upstream designated ($O_3$) and 401 bp downstream ($O_2$) of *lac* operator. My student Monika Besse then showed that the first operator could be moved away from the promoter without reducing repression [14], and, later, another student of my group, Elisabeth Eismann, showed that destruction of the second operator within the *lac* Z gene decreased repression three to five fold in an episomal system [15]. Other laboratories have reported similar results using different techniques.

The main problem with this type of analysis was that it was done on plasmids with high amounts of *lac* repressor rather than on the chromosome. Moreover, controls with dimeric *lac* repressor were lacking. A group of graduate students (among them Jürgen Kun and Klaus Sieg) recently constructed a *Lambda* vector system, which allows the rapid introduction of *in vitro* manipulated *lac* systems into the bacterial chromosome [16]. With this system Oehler showed that indeed the first operator alone represses only 17 fold [17]. The rest of the thousand fold repression is brought about by the second and the third operators. Oehler showed in addition that these two auxiliary operators do not (or only marginally) mediate repression by themselves. They function primarily by helping the first operator in repression. This changes our view of repression in a most unexpected manner. Pseudo-operators become auxiliary operators [17]. These *in vivo* experiments confirm earlier *in vitro* experiments made by my graduate students Helmut Krämer and Elisabeth Eismann in collaboration with Michele Amouyal (Institut Pasteur, Paris) and Albert Nordheim (ZMBH, Heidelberg). They showed that in linear [18] and circular [19, 20] DNA, loops between two *lac* operators and a tetrameric *lac* repressor occur and, in fact, are particularly strong compared to the single occupation of one particular operator.

## 2.1.2.4   Recognition of *lac* Operator by *lac* Repressor

When I came to Köln in 1968, I knew a way to elucidate which part of *lac* repressor recognized *lac* operator. At Harvard, I had isolated a special class of *lac* I gene mutants, which were negative dominant ($I^{-d}$) [3]. Such mutant *lac* repressor were tetrameric, bound inducer *in vitro*, but could not bind operator *in vitro* or *in vivo* (i. e., it could not repress). My students Bruno Gronenborn and Magnus Pfahl mapped 200 such $I^{-d}$ mutants and found that they all mapped at the extreme 5' end of the *lac* I gene [21]. I may add here that Pfahl showed that the $I^s$ mutations, which destroyed inducer binding but not operator binding, were scattered throughout the *lac* I gene [22]. Jeffrey Miller later provided a much more detailed map of these mutations [5]. Extrapolating from the $I^{-d}$ mapping data to the *lac* repressor protein, we came to the conclusion that the N-terminal 50 residues were sufficient for operator recognition [23]. This was confirmed later when another graduate student, Michael Koenen, showed that 76 N-terminal residues of homologous *gal* repressor was sufficient when fused to β-galactosidase to convert β-galactosidase into a repressor of the *gal* system [24].

It should be recalled that wild type *lac* operator is rather asymmetric. Thus, for example, Walter Gilbert concluded that its recognition might be complicated; different subunits might recognize different DNA sequences. To solve this riddle, we set out to isolate the simplest possible *lac* operator from a random collection of DNA. We assumed that the introns in mouse DNA qualified as being random. We, therefore searched mouse DNA for *lac* operator-like structures, i.e. structures which bound specifically *lac* repressor. Anne Marie Simons, a student in my lab, was successful and isolated several variants of what we now call ideal, symmetric *lac* operator, a totally symmetric *lac* operator, which lacks the central base pair of natural operator [25].

After the discovery of ideal *lac* operators and the advent of techniques of automated DNA synthesis, we could begin an extensive investigation of the molecular details of the recognition of *lac* repressor by *lac* operator. We had speculated as early as 1973 that a putative alpha helix, starting with residue 17 and ending with residue 26, would lie in the major groove of *lac* operator DNA. We further speculated that this recognition helix would recognize the particular sequence and that this type of interaction between an alpha helix and the major groove of DNA would serve as a general mechanism of recognition and furthermore, that general rules (a code) may exist, which would govern this recognition [23].

We were pleasantly surprised when Brian Matthews showed that *lambda cro* protein, indeed, has such a recognition helix, that a homologous helix exists in *lac* repressor, and that indeed the putative recognition helix of *lac* repressor starts with residue 17, as we had predicted. Brigitte von Wilken-Bergmann constructed a plasmid system, which allows the fast testing by an *in vivo* repression test of repressor-operator interactions [26]. The system consists of a bacterial host, which carries a *lac* deletion and two plasmids with different origins of replication and different antibiotic resistance markers. One plasmid carries a *lac I* gene whose 5′ end we had replaced by synthetic DNA carrying as many unique restriction sites as possible. In this plasmid, we could create any mutant or bank of mutants at the 5' end of the *lac I* gene. The other plasmid carries a *lac* operon with a wild type *lac* promoter and a deletion at the site where *lac* operator is normally positioned. We inserted an unique restriction site in the place of this deletion in such a way that we could easily insert any synthetic operator DNA.

We tested first all possible symmetric variants of ideal symmetric *lac* operator beginning at base pair 1 (next to the axis of symmetry) and ending at base pair 9 (9 base pairs away from the axis of symmetry). We tested all these constructs, including the construct with the ideal symmetric operator with wild type *lac* repressor, and found that ideal symmetric *lac* operator does, indeed, lead to the highest repression value. We concluded that ideal symmetric operator is, in fact, the best possible operator for *lac* repressor [26].

Two graduate students in the lab, Norbert Lehming and Jürgen Sartorius, then set out to analyse which residues of this recognition helix interact with which base pairs of the *lac* operator. We decided to take only strong changes of specificity as indicating the interaction of a particular side chain with a particular base pair. Such a mutant repressor had to repress ideal *lac* operator to a lesser degree than wild type repressor. At the same time, it had to repress a particular symmetric mutant operator better than the ideal *lac* operator.

We used two strategies to find such changes of specificity. First, we looked at homologous repressors and operators and tried to introduce minimal changes into the *lac* system. For example, *gal* repressor had a Val1Ala2 sequence in the first two residues of its recognition helix, in contrast to *lac* repressor, which had a Tyr1Gln2 sequence. On the other hand, the DNA sequence of the *gal* operator had an A in position 4 where *lac* operator had a G. We then replaced the Tyr1Gln2 sequence in the recognition helix of *lac* repressor by Val-Ala and tested it with a *lac* operator, in which base pair 4 had been replaced by A, and indeed, we found a specificity change [26].

The second strategy was to use a small banks. Here, not just one mutant but a set of mutants could be tested with various operators. In this type of approach, we found, for example, mutant *lac* repressors in which Tyr 1 was replaced by His to bind strongly to a *lac* operator variant, which had the G in base pair 4 replaced by a T.

At the moment, we have finished an exhaustive analysis of amino acid residues 1, 2, and 6 of the recognition helix of *lac* repressor. We have shown that the actions of residues 1, 2, and 6 are additive and that the recognition helix of *lac* repressor is oriented oppositely to that of *lambda* repressor. Unpublished evidence of Sartorius indicates that residue 5 of the recognition helix binds to base pair 4. An analysis of residue 9 is underway. A comparison with various other bacterial repressors indicates that there exists a group of repressors, which uses similar, if not the same, rules of base pair recognition, as does *lac* repressor [27– 29].

### 2.1.2.5   Replacing *lac* Operator by Other Targets: The Targets of *trp* Repressor, *gal4* Protein and Putative Poly $GC_n$ Binding Protein

The plasmids we described above can be used for the analysis of other systems, when one replaces the *lac* operator by the target one wants to analyse. I had always followed with interest the analysis of the *trp* system. In the first year of the Sonderforschungsbereich 74, Jürgen Schrenk, a graduate student, and Jeffrey Miller, then a postdoctoral fellow in my group, isolated an *in vivo* fusion between the *trp* promoter operator-region and the *lac Z* gene. This construct was used by Geoffrey Zubay (Columbia University, New York) for the detection and isolation of *trp* repressor [30]. My students Barbara Walter and Detlef Staake recently inserted synthetic *trp* operator in place of the *lac* operator and measured repression of such constructs by *trp* repressor. The main result of this analysis is that the generally accepted axis of symmetry of *trp* operator between one *trp* repressor dimer is wrong. Our results question the interpretations of recent x-ray work of Paul Sigler [31]. Since his oligonucleotide carried the wrong axis of symmetry his repressor operator complex was nonspecific. All conclusions about a new mode of repressor operator recognition are therefore doubtful.

Together with Moshe Yaniv (Institut Pasteur, Paris) we showed that the *yeast ga14* protein represses β-galactosidase synthesis, when *lac* operator is replaced by the *ga14* target [32]. Other attempts by us to show in *E. coli* similar effects by finger proteins or homeobox proteins have so far not given positive results. We have used the system, however, to analyse successfully the effect of Z-DNA in the form of (GC) oligomers inserted in place of *lac* operator. We showed that expression of β-galactosidase goes down with increasing length of the GC units. Furthermore, we showed that this effect is not due to the existence of protein binding to (GC) and that such constructs are unstable and loose GC units [33].

## 2.1.3   The *lac* Z (β-galactosidase) Gene

### 2.1.3.1   Structure of the *lac Z*, the *gal ET* Genes of *E. coli*

Irvin Zabin determined the amino acid sequence of β-galactosidase, the largest protein ever sequenced. To confirm it Angela Kalnin set out to sequence the *lac Z* gene [34]. Nothing special could be seen; the protein sequence was largely correct, but it was certainly useful to know the nucleotide sequence for all further construction work. I would like to mention here one particular construct: Wolfgang Kuchinke, a graduate student, fused two *lac Z* genes in phase head to tail and found an active, more heat stable β-galactosidase of the same molecular weight as the wild-type enzyme [35]. I may just note here that our expectation of seeing homologies among the three *gal* enzymes of *E. coli* (galactose kinase, UDPG-transferase, epimerase) were not born out [36]. We constructed over-producers and gave them to Keith Wilson (EMBL, Hamburg) for possible X-ray analysis, expecting similar three dimensional structures.

## 2.1.3.2   The *lac* Z Gene as Part of Cloning Vectors

Last but not least, the *lac* system became a paradigm of gene regulation because the product of its most prominent structural gene, β-galactosidase, could be easily tested by a sensitive colorimetric test. So β-galactosidase became an ideal reporter enzyme. After I had isolated the first functional β-galactosidase fusion in 1974 [9] the *lac* Z gene was developed by Jon Beckwith and his collaborators (Harvard Medical School, Boston, Mass., USA) to express β-galactosidase as reporter enzyme in an *in vivo* transposition system using phage *mu*.

Bruno Gronenborn, working as a postdoctoral fellow in my group, and Joachim Messing, from Peter-Hans Hofschneider's laboratory (MPI, München-Martinsried), combined some of the advantages of the *lac* Z system with the advantages of phage *M13*. They constructed a derivative of phage *M13*, which carried the *lac* promoter and the first 60 condons of the *lac* Z gene [37]. The replicative form of this *M13* derivative thus produces mRNA coding for a N-terminal fragment of β-galactosidase, which is able to complement an inactive mutant β-galactosidase produced from a *lac* Z gene, which carries a deletion of codon 11–41 (termed alpha complementation). Cloning into this *lac* Z fragment destroys the capacity to complement. This lack of complementation can easily be screened for. Gronenborn participated in the first steps to convert the original construct into a cloning vector by introducing a unique restriction site into the Z fragment on *M13* [38]. A similar plasmid vector was constructed by my student Uli Rüther in pBR322 [39]. The later development of M13 vector was then done by Messing and others. Now derivatives of *M13 lac* are the most used cloning vectors for dideoxy DNA sequencing.

My main interest went into vectors in which active β-galactosidase fusions were produced. Uli Rüther constructed such plasmid vectors carrying a *lac* Z gene with unique restriction sites at the extreme 5' [40, 41] or 3' [42] end. Suitably inserted genomic DNA is expressed as active β-galactosidase fusion protein. Particular fusions can be detected in colonies by suitable antibodies. The fusion proteins can be purified on thiogalactoside gels. Rüther tried the method with a clone of genomic DNA containing exons 1 and 2 of chicken lysozyme. He succeeded in detecting such clones carrying epitopes of these two exons in a library of about 10 000 clones [41].

Then I decided to use this method for the possible isolation of clones one might use in the vaccination against a particular parasite. My graduate student Michael Koenen constructed a library of genomic *Plasmodium falciparum* DNA and screened it with human sera of malaria immune donors [43]. The screen of this library revealed 30 different clones. One of them belonged to an unknown new class of repetitive DNA not found in cDNA libraries [44]. Jürgen Kun, another graduate student, used the analogous *Lambda lac* Z expression vector for expression cloning of genomic *P.*

*falciparum* DNA. He constructed a *Lambda* library which was about 10 times larger than the previous plasmid library. He screened this bank with antiserum prepared against *P. falciparum* membranes prepared by Michael Schreiber. He found several asparagine rich, partially repetitive clones. Jürg Gysin (Institut Pasteur, Cayenne) found that the purified fusionproteins inhibited opsonisation of *Plasmodium* infected erythrocyte activated by protective monkey sera. Antibodies against these clones indicated that the antigen was present on the membrane of parasite infected erythrocytes. These antigens are at the moment being tested by Gysin in *Saimiri* monkeys as possible vaccines. Finally I would like to mention that we have used our vectors to construct and express bovine trypsin inhibitor from synthetic DNA [45].

## 2.1.4 Structure and Functional Analysis of *lac* Permease

The product of the *lac Y* gene *lac* permease, the second major structural protein from the *lac* system of *E. coli*, is most interesting from a functional point of view: it transports β-galactosides and $H^+$ ions against a gradient into the cell. Ann Hobson, a postdoctoral fellow, did a deletion mapping of a large set of *Y* nonsense mutants [46]. My graduate student Dagmar Büchel sequenced the *lac Y* gene. The sequence showed that *lac* permease contains 12 long lipophilic regions, presumably alpha helices, spanning the inner *E. coli* membrane [47]. Further analysis by my two graduate students Hans Werner Griesser and Hubertus Bocklage, and Peter Overath (MPI Tübingen) indicated that the extreme N-terminus of *lac* permease can be replaced by other sequences [48, 49] and that exchanges in the first membrane spanning alpha helix destroy proton but not galactoside transport [50]. Martin Mieschendahl showed, furthermore, that mutants, which transported lactose only at high concentrations, clustered in particular regions of the *Y* gene [51]. His further suggestion of the existence of negative dominant mutants could not be verified *in vitro*: *lac* permease is a monomer as extensive investigation by Overath and collaborators has shown [52]. Finally, my student Martina Markgraf isolated and sequenced a mutant, which transports maltose better than lactose and, thus, defines part of the region coding for sugar specificity [53].

## 2.1.5 Amyloid Precursor – Leaving the *lac* System for an Excursion into the Molecular Biology of Alzheimer's Disease

My former collaborator, Konrad Beyreuther, then an associate professor (C3 Professor) in this institute, had turned his interest to Alzheimer's disease and particularly to the amyloid found in the brains of Alzheimer patients. The question arose whether amyloid was coded by a human gene or possibly by a virus. At that time, he had just one student, Michael Salbaum, searching genomic libraries for the gene. A graduate student of mine, Axel Unterbeck, had prepared a cDNA bank of human embryonal brain mRNA. Another student, Jie Kang, screened this bank successfully for a cDNA coding for amyloid. The cDNA coding for 695 residues was sequenced by another graduate student, Hans Georg Lemaire [54]. The deduced protein sequence looked like a receptor protein. Salbaum showed that the gene was coded on chromosome 21 [54]. Beyreuther and his collaborators demonstrated later that the protein indeed behaved *in vitro* like a receptor protein [55]. Kang and Lemaire together with collaborators of Beyreuther (now ZMBH Heidelberg) and Unterbeck (now Bayer, New Haven) later sequenced the intron-exon boundaries of this gene [56 see also 57]. Others found splice products, which contained two additional exons among them, one resembling trpsin inhibitor. Kang recently used PCR for quantitative determinations. She showed that the form lacking these two extra exons is about 40 times over-represented in the brain but occurs in about equal amounts in all other tissues [58]. At the moment, Kang is trying to clone brain proteases, which may possibly be involved in the degradation of amyloid precursor.

## 2.1.6 Beyond Molecular Biology: Excursions into Philosophy and History of Science

This report would be incomplete if I did not mention that I have directed my effort from time to time to problems which are rather outside molecular biology. I can say, however, that without working constantly on problems of molecular biology, I would not have had the necessary scientific perspective on those other problems.

I wrote two books. One (*"Die Philosophen und das Lebendige"*) deals with two particular aspects of philosophy of science i. e. how various philosophers looked at living matter and whether it is worthwhile to deal with dialectical materialism [59]. I may just say here that I still find both questions most stimulating. The book raised little professional interest, although it was translated into Italian. The other book ("*Tödliche Wissenschaft*" or "Murderous Science") deals with the history of human genetics in Nazi-Germany [60]. It did not at first raise much interest in this country either but some interest elsewhere. It was translated into English, French, Spanish, Italian, and Dutch. A translation into Hebrew is in preparation. I showed here, among other things, that the Deutsche Forschungsgemeinschaft supported the genocide of the Gypsies and the research Josef Mengele did in Auschwitz for his advisor Otmar von Verschuer of the Kaiser Wilhelm-Institut für Anthropologie in Berlin-Dahlem. A follow-up article [61] was translated into French. Another one was written for a textbook on psychiatric ethics [62]. I may also add that I reviewed some books for Nature and other journals and wrote some literary articles of which I would like to reference just one [63]. It deals with my own dreams, the very other end of science.

At the moment when I am writing these lines it seems as if misguided public opinion may lead the political parties and courts to stop DNA-manipulation altogether in the FRG. I would like to document here that I wrote against this tide of unreason [64, 65]. I may conclude by saying, that I am immensely grateful to the Deutsche Forschungsgemeinschaft and other sponsors, who have supported my scientific work for so many years in spite of all possible irritation my excursions into different fields may have caused temporarily.

## 2.1.7   Summary and Outlook

The techniques of cloning, sequencing, and synthesis of DNA drove most molecular biologists, who had worked previously with *E. coli*, into the virgin fields of eucaryotes. I stayed with *E. coli* and with the system I had begun to work with, the *lac* system. I hope that this report makes clear that this was a good decision. The new knowledge gained was not simply more of the same but fundamental knowledge at a deeper level of understanding. We begin now to understand how *lac* repressor regulates the *lac* system by forming loops with two operators. Repressor-operator recognition and repression can be now understood at the molecular level. The molecular mechanisms used in the *lac* system are generally used. And finally the *lac* Z (β-galactosidase) gene was converted into a tool in a series of widely used vectors. Gunther Stent's Golden Age turned out to be the Stone Age of molecular biology. I am pretty confident in predicting that fundamental problems still wait to be solved just in procaryotes. For those who are curious because they realize what they do not know, the quest for knowledge never ends.

Benno Müller-Hill

*Acknowledgements*

Papers marked (A) have been supported by the Deutsche Forschungsgemeinschaft through Sonderforschungsbereich 74, those marked (B) by the Deutsche Forschungsgemeinschaft through Sonderforschungsbereich 234, those marked (C) by the Deutsche Forschungsgemeinschaft through other programms, those marked (D) by Stiftung Volkswagen, those marked (E) by Thyssen-Stiftung, those marked (F) by BMFT, those marked (G) by Fonds der Chemie, those marked (H) by other sponsors. Papers marked (J) had no sponsors but belong to the publications of Benno Müller-Hill and his group. Generous ground support by the Land Nordrhein-Westfalen is here gratefully acknowledged.

## 2.1.8   References

[1]  A. Lwoff, A. Ullmann (eds.): Selected Papers in Molecular Biology by Jacques Monod. Academic Press, New York 1978.
[2]  W. Gilbert, B. Müller-Hill: Isolation of the *lac* repressor. Proc. Natl. Acad. Sci. USA 56 (1967) 1891–1898.
[3]  B. Müller-Hill, L. Crapo, W. Gilbert: Mutants that make more *lac* repressor. Proc. Natl. Acad. Sci. USA 59 (1968) 1259–1264.
[4]  J. R. Beckwith, D. Zipser (eds.): The Lactose Operon. Cold Spring Harbor Laboratory, Cold Spring Harbor, N. Y. 1970.
[5]  J. H. Miller, W. S. Reznikoff (eds.): The Operon. Cold Spring Harbor Laboratory, Cold Spring Harbor, N. Y. 1978.
[6]  K. Beyreuther, K. Adler, N. Geisler, A. Klemm: The Amino Acid Sequence of *lac* repressor. Proc. Natl. Acad. Sci. USA 70 (1973) 3567–2576 (A, D).
[7]  B. v. Wilcken-Bergmann, B. Müller-Hill: Sequence of *gal R* Gene Indicates a Common Evolutionary Origin of *lac* and *gal* Repressor in *Escherichia coli*. Proc. Natl. Acad. Sci. USA 79 (1982) 2427–2431. (A)
[8]  B. Müller-Hill: Sequence Homology between *lac* and *gal* Repressors and three Sugar-Binding Periplasmic Proteins. Nature 302 (1983) 163–164. (G)
[9]  B. Müller-Hill, J. Kania: *Lac* Repressor can be fused to β-galactosidase. Nature 249 (1974) 561–563. (A)
[10] A. J. Brake, A. V. Fowler, I. Zabin, J. Kania, B. Müller-Hill: β-galactosidase chimeras: Primary Structure of *lac* Repressor-β-Galactosidase Protein. Proc. Natl. Acad. Sci. USA 75 (1978) 4824–4827. (A)
[11] J. Kania, D. T. Brown: The functional repressor parts of a tetrameric *lac* repressor-β-galactosidase chimaera are organized as dimers. Proc. Natl. Acad. Sci. USA 73 (1976) 3529–3532. (A)
[12] J. Kania, B. Müller-Hill: Construction, Isolation and Implications of Repressor-β-galactosidase Hybrid Molecules. Eur. J. Biochem. 79 (1977) 381–386. (A)
[13] S. Alberti, S. Oehler, B. Müller-Hill: The C-terminus of *lac* repressor resembles a leucine zipper and is able to make dimers of *lac* repressor aggregate to tetramers. The New Biologist 3 (1991) 57–62. (B)

38

[14] M. Besse, B. v. Wilcken-Bergmann, B. Müller-Hill: Synthetic *lac* Operator Mediates Repression Through *lac* Repressor when introduced Upstream and Downstream from *lac* Promotor. EMBO J. 5 (1986) 1377–1381. (A)

[15] E. Eismann, B. v. Wilcken-Bergmann, B. Müller-Hill: Specific destruction of the second *lac* operator decreases repression of the *lac* operon in *Escherichia coli* fivefold. J. Biol. 195 (1987) 949–952. (A)

[16] K. Sieg, J. Kun, A. Scherf, B. Müller-Hill: A versatile vector system for expression cloning in *Escherichia coli*. Gene 75 (1989) 261–270. (C)

[17] S. Oehler, E. Eismann, H. Krämer, B. Müller-Hill: The three operators of the *lac* operon cooperate in repression. EMBO J. 9 (1990) 973–979.(B)

[18] H. Krämer, M. Niemöller, M. Amouyal, B. Revet, B. v. Wilcken-Bergmann, B. Müller-Hill: *lac* repressor forms loops with linear DNA carrying two suitably spaced *lac* operators. EMBO J. 6 (1987) 1481–1491. (A)

[19] H. Krämer, M. Amouyal, A. Nordheim, B. Müller-Hill: DNA supercoiling changes the spacing requirement of two *lac* operators for DNA loop formation with *lac* repressor. EMBO J. 7 (1988) 547–556. (A, F)

[20] E. Eismann, B. Müller-Hill: *lac* repressor forms stable loops *in vitro* with supercoiled wild type *lac* DNA containing all three natural operators. J. Mol. Biol. 213 (1990) 763–775. (B)

[21] M. Pfahl, C. Stockter, B. Gronenborn: Genetic analysis of the active sites of *lac* repressor. Genetics 76 (1974) 669–679. (A)

[22] M. Pfahl: Genetic map of the lactose repressor gene (i) of *Escherichia coli*. Genetics 72 (1972) 393–410. (A)

[23] K. Adler, K. Beyreuther, E. Fanning, N. Geisler, B. Gronenborn, A. Klemm, B. Müller-Hill, M. Pfahl, A. Schmitz: How *lac* repressor binds to DNA. Nature 237 (1972) 322–327. (A)

[24] B. v. Wilcken-Bergmann, M. Koenen, W. Griesser, B. Müller-Hill: 72 Residues of *gal* repressor fused to β-galactosidase repress the *gal* operon of *E. coli*. EMBO J. 2 (1983) 1271–1274. (A)

[25] A. Simons, D. Tils, B. v. Wilcken-Bergmann, B. Müller-Hill: Possible ideal *lac* operator: *E. coli lac* operator-like sequences from eukaryotic genomes lack the central GC pair. Proc. Natl. Acad. Sci. USA 81 (1984) 1624–1629. (A)

[26] N. Lehming, J. Sartorius, M. Niemöller, G. Genenger, B. v. Wilcken-Bergmann, B. Müller-Hill: The interaction of the recognition helix of *lac* repressor with *lac* operator. EMBO J. 6 (1987) 3145–3153. (A, F)

[27] N. Lehming, J. Sartorius, S. Oehler, B. v. Wilcken-Bergmann, B. Müller-Hill: The recognition helices of *lac* and *lambda* repressor are oriented in opposite directions and recognize similar DNA sequences. Proc. Natl. Acad. Sci. USA 85 (1988) 7947–7951. (A)

[28] J. Sartorius, N. Lehmig, B. Kisters, B. v. Wilcken-Bergmann, B. Müller-Hill: *lac* repressor mutants with double or triple exchanges in the recognition helix bind specifically to *lac* operator variants with multiple exchanges. EMBO J. 8 (1989) 1265–1270. (A, F)

[29] N. Lehming, J. Sartorius, B. Kisters-Woike, B. v. Wilcken-Bergmann, B. Müller-Hill: Mutant *lac* repressors with new specificities hint at rules for protein DNA recognition. EMBO J. 9 (1990) 615–621 (B, F)

[30] J. Zubay, D. Morse, W. J. Schrenk, J. H. M. Miller: Detection and isolation of the repressor protein for tryptophan operon of *Escherichia coli*. Proc. Natl. Acad. Sci. USA 69 (1972) 1100–1103. (A)

[31] B. Walter, T. Staake, B. Müller-Hill: How *trp* repressor binds to its target *trp* operator. EMBO J. 9 (1990) 1963–1967 and Corrigendum 9 (1990) 3023. (B, F)

[32] N. Paulmier, M. Yaniv, B. v. Wilcken-Bergmann, B. Müller-Hill: *Gal4* transcription activator protein of yeast can function as a repressor in *E. coli*. EMBO J. 6 (1987) 3539–3542. (A)

[33] E. Horbach, B. Müller-Hill: Insertion of $d(pCpG)_n \cdot d(pCpG)_n$ into the *lac Z* gene of *E. coli* inhibits expression of β-galactosidase *in vivo*. J. Mol. Biol. 202 (1988) 157–160. (A)

[34] A. Kalnins, K. Otto, U. Rüther, B. Müller-Hill: Sequence of the *lac Z* Gene of *E. coli*. EMBO J. 2 (1983) 593–597. (A)

[35] W. Kuchinke, B. Müller-Hill: Fused *lac Z* Genes code for di-, tri- and tetra-β-galactosidase in *E. coli*. EMBO J. 4 (1985) 1067–1073. (A)

[36] H.-G. Lemaire, B. Müller-Hill: Nucleotide sequences of the *gal E* and *gal T* genes of *E. coli*. Nucl. Acid. Res. 14 (1986) 7705–7711. (A)

[37] J. Messing, B. Gronenborn, B. Müller-Hill, P. Hofschneider: Filamentous coliphage *M13* as a cloning vehicle. Proc. Natl. Acad. Sci. USA 74 (1977) 3642–3646. (A)

[38] B. Gronenborn, J. Messing: Methylation of single stranded DNA in vitro introduces new restriction endonuclease cleavage sites. Nature 272 (1978) 375–377. (A)

[39] U. Rüther, M. Koenen, K. Otto, B. Müller-Hill: pUR222, a vector for cloning and rapid chemical sequencing of DNA, Nucl. Acids. Res. 9 (1981) 4087–4098. (C)

[40] M. Koenen, U. Rüther, B. Müller-Hill: Immunoenzymatic detection of expressed gene fragments cloned in the *lac Z* gene of *E. coli*. EMBO J. 4 (1982) 509–512. (C)

[41] U. Rüther, M. Koenen, A. Sippel, B. Müller-Hill: Exon cloning: Immunoenzymatic identification of exons of the chicken lysozyme gene. Proc. Natl. Acad. Sci. USA 79 (1982) 6852–6855. (C)

[42] U. Rüther, B. Müller-Hill: Easy Identification of cDNA clones. EMBO J. 2 (1983) 1791–1794. (C)

[43] M. Koenen, A. Scherf, O. Mercereau, G. Langsley, L. Sibilli, P. Dubois, L. Pereira da Silva, B. Müller-Hill: Human antisera detect a *Plasmodium falciparum* genomic clone encoding a nonapeptide repeat. Nature 311 (1984) 382–384. (C)

[44] A. Scherf, C. Hilbich, K. Sieg, D. Mattei, O. Mercereau-Puijalon, B. Müller-Hill: The 11-1 gene of *Plasmodium falciparum* codes for distinct fast evolving repeats. EMBO J. 7 (1988) 1129–1137. (C)

[45] B. v. Wilcken-Bergmann, D. Tils, J. Sartorius, E. A. Auerswald, W. Schröder, B. Müller-Hill: A synthetic operon containing 14 bovine pancreatic Trypsin inhibitor genes is expressed in *E. coli*. EMBO J. 5 (1986) 3219–3225. (F, H)

[46] A. C Hobson, D. Gho, B. Müller-Hill: Isolation, Genetic Analysis and Characterisation of *Escherichia coli* mutants with Defects in the *lac Y* Gene. I. Bact. 131 (1977) 830–838. (A)

[47] D. E. Büchel, G. Gronenborn, N. Müller-Hill: Sequence of the lactose permease gene. Nature 283 (1980) 541–545. (A)

[48] H. Bocklage, B. Müller-Hill: *lac Z⁻ – Y⁺* fusions in *E. coli*. DNA sequencing reveals the eight N-terminal residues of *lac* permease as Non-essential. Eur. J. Biochem. 137 (1983) 561–565. (A)

[49] H. W. Griesser, B. Müller-Hill, P. Overath: Characterisation of β-galactosidase-lactose-permease chimeras in *E. coli*. Eur. J. Biochem. 137 (1983) 567–572. (A)

[50] M. Mieschendahl, D. Büchel, H. Bocklage, B. Müller-Hill: Mutations in the *lac Y* gene of *E. coli* define functional organization of lactose permease. Proc. Natl. Acad. Sci. USA 78 (1981) 7652–7656. (A)

[51] P. Overath, U. Weigel, J.-M. Neuhaus, J. Soppa, R. Seckler, I. Riede, H. Bocklage, B. Müller-Hill, G. Aichele, K. Wright: Lactose permease of *E. coli*: Properties of mutants defective in substrate translocation. Proc. Natl. Acad. Sci. USA 84 (1987) 5535–5539. (A)

[52] K. Wright, U. Weigel, H. Bocklage, M. Mieschendahl, B. Müller-Hill, P. Overath: Does the lactose carrier of *E. coli* function as a monomer? FEBS Lett. 162 (1983) 11–15. (A)

[53] M. Markgraf, H. Bocklage, B. Müller-Hill: Change of Threonine 266 to Isoleucine in *lac* permease of *E. coli* diminishes the transport of lactose and increase the transport of Maltose. Mol. Gen. Genetics 198 (1985) 473–475. (A)

[54] J. Kang, H.-G. Lemaire, A. Unterbeck, J. M. Salbaum, C. L. Masters, K. H. Grzeschick, G. Multhaup, K. Beyreuther, B. Müller-Hill: The precursor of Alzheimer's Disease Amyloid A4 protein resembles a cell surface receptor. Nature 325 (1987) 733–736. (A)

[55] T. Dyrks, A. Weidemann, G. Multhaup, J. M. Salbaum, H.-G. Lemaire, J. Kang, B. Müller-Hill, C. L. Masters, K. Beyreuther: Identification, transmembrane orientation and biogenesis of the amyloid A4 precursor of Alzheimer's disease. EMBO J. 7 (1988) 949–957. (E)

[56] H.-G. Lemaire, J. M. Salbaum, G. Multhaup, J. Kang, R. M. Bayney, A. Unterbeck, K. Beyreuther, B. Müller-Hill: The PreA4$_{695}$ precursor protein of Alzheimer's disease A4 amyloid is encoded by 16 exons. Nucl. Acid. Res. 17 (1989) 517–522. (E)

[57] B. Müller-Hill, K. Beyreuther: Molecular Biology of Alzheimer's disease. Annual Reviews of Biochemistry 58 (1989) 287–307. (E)

[58] J. Kang, B. Müller-Hill: Differential splicing of Alzheimer's disease amyloid A4 precursor RNA in rat tissues: PreA4$_{695}$ mRNA is predominantly produced in rat and human brain. Biochem. Biophys. Res. Comm. 166 (1990) 1192–1200. (E)

[59] B. Müller-Hill: Die Philosophen und das Lebendige. Campus, Frankfurt 1981. (I)

[60] B. Müller-Hill: Tödliche Wissenschaft. Die Aussonderung von Juden, Zigeunern und Geisteskranken 1933–1945. Rowohlt, Reinbeck 1984; English translation by G. Fraser: Murderous Science. Elimination by Scientific Selection of Jews, Gypsies and Others in Germany 1933–1945. Oxford University Press, Oxford 1988. (I)

[61] B. Müller-Hill: Genetics after Auschwitz. Holocaust and Genocide Studies 2 (1987) 3–20. French translation: Temps Mod. 511 (1989) 52–85. (I)

[62] B. Müller-Hill: The lowest point in the history of psychiatry: Nazi Germany. In: S. Bloch, P. Chodoff (eds.): Psychiatric Ethics. Oxford University Press, 2nd edn., (1991) 461–472. (I)

[63] B. Müller-Hill: Nachts, Sinn und Form 41 (1989) 637–650. (I)

[64] B. Müller-Hill: Gentechnologie als Beruf, TAZ 1.2.1989. (I)

[65] B. Müller-Hill: Schluß mit der Gentechnologie oder Science Fiktion in der BRD. Forum Wissenschaft 6 (1989) 7–11. (I)

# Former and Present Collaborators

## Technical Assistant

Ellen Fanning: Professor, Universität München, FRG.

## Graduate Students

Klaus Adler: Medical student.
Monika Besse: Private life.
Hubertus Bocklage: Postdoctoral fellow, Universität Würzburg, FRG.
Dagmar Büchel: Private life.
Elis Eismann: Postdoctoral fellow, Kernforschungsanlage Jülich, FRG.
Norbert Geisler: Research fellow, MPI für Biophysikalische Chemie Göttingen, FRG.
Hans Werner Griesser: Industry, FRG.
Bruno Gronenborn: Directeur de recherche, CNRS Gif sur Yvette, France.
Dietmar Kamp: Professor, University of Bergen, Norvegia.
Jie Kang: Industry, FRG.
Jürgen Kania: Medical student.
Alex Klemm: Death in an accident, 1972.
Michael Koenen: Postdoctoral fellow, Medizinische Hochschule Hannover, FRG.
Helmut Krämer: Postdoctoral fellow, UCLA, Los Angeles, USA.
Wolfgang Kuchinke: Postdoctoral fellow, Osaka Bioscience Institute, Japan.
Norbert Lehming: Postdoctoral fellow, Harvard University, USA.
Hans-Georg Lemaire: Industry, FRG.
Martin Mieschendahl: Industry, FRG.
Uli Rüther: Research fellow, EMBL Heidelberg, FRG.
Jürgen Sartorius: Industry, FRG.
Arthur Scherf: Chercheur, Institut Pasteur, Paris, France.
Martin Schlotmann: Medical student.
Albert Schmitz: Industry, Switzerland.
Michael Schreiber: Postdoctoral fellow, Tropeninstitut, Hamburg, FRG.
Peter Schreier: Industry, FRG.
Jürgen Schrenk: Industry, FRG.
Klaus Sieg: Unknown.
Axel Unterbeck: Industry, FRG.
Brigitte von Wilcken-Bergmann: Research fellow in the group.

## Postdoctoral Scientists

Konrad Beyreuther: Professor, Universität Heidelberg, FRG.
John Betteridge: Industry, UK.
Ann Hobson: Postdoctoral fellow, University of California, Berkely, USA.
Tom Fanning: Researcher, NIH, Washington, USA.
Gustavo Maroni: Assistant Professor, University of North Carolina Chapel Hill, USA.
Jeffrey Miller: Professor, UCLA, Los Angeles, USA.

## Guest Scientists

Robert Beabealashivili, Academy of Medical Sciences, Moscow, USSR.
Sergei Grokowsky: Academy of Sciences, Moscow, USSR.
El. Abidim A.Z. Salam: Professor, Ains Shams University, Cairo, Egypt.
Nguyen Quang Vinh: National Scientific Research Center, Hanoi, Vietnam.

## Habilitation

Konrad Beyreuther: Professor, Universität Heidelberg, FRG.

# 2.2    Of Lipids and Membranes*

Peter Overath**

In 1969, the Deutsche Forschungsgemeinschaft established a research group working on the "Genetics of Bacterial Lipid and Membrane Metabolism" at the Institut für Genetik, Universität zu Köln. This group was subsequently integrated into the Sonderforschungsbereich "Molekularbiologie der Zelle" until 1973, when it moved to the Max-Planck-Institut für Biologie in Tübingen.

Three projects were persued using *Escherichia coli* as a model organism: The first was the characterization of the genes coding for and regulating the synthesis of the enzymes of fatty acid degradation [1, 2]. These studies laid the ground work for a more detailed analysis using molecular biological techniques in other laboratories (cf. Ref. [3] for a recent review). The second was the isolation and biochemical analysis of mutants of the ATPase complex and the respiratory chain [4, 5]. The goal of the third project was the isolation of mutants defective in the synthesis of fatty acids and their use in the manipulation of membrane phospholipid composition. I will restrict the discussion to the latter project because it leads to conclusions of more general significance.

The membranes of both eukaryotic and prokaryotic organisms have a complex lipid composition regarding both the apolar fatty acyl chains and the polar head groups. The question was and still is today how much of this complexity is required for the assembly and function of membranes. Function of phospholipids can be considered on two levels. On the one hand, the composition of phospholipids must secure the formation of an efficient barrier separating two aqueous compartments. This role mainly depends on the physical properties of the molecules, such as the dimensions of the polar head groups and hydrocarbon chains, rather than details of their chemical structure. On the other hand, certain phospholipids can interact very specifi-

---

* This article is dedicated in gratitude to Dr. Dr. h.c. Anita Hoffmann
** Institut für Genetik der Universität zu Köln
    (at present: Max-Planck-Institut für Biologie, Tübingen)

44

cally with defined proteins and, therefore, have a cofactor-like function. Examples for such specific interactions have remained scarce. Our experiments concerned the relationship between phospholipid structure and the bulk physical properties of membranes.

Together with H. U. Schairer, mutants were isolated, which were defective in the synthesis of unsaturated fatty acids carrying a *cis*-double bond [6, 7]. These mutants could be supplemented with fatty acids normally not present in *E. coli* phospholipids. In particular, fatty acids carrying a *trans*-double were incorporated to a high extent resulting in an essentially homogenous phospholipid hydrocarbon chain composition. The physiology of these cells as well as the physical properties of derived membranes and lipids were studied in detail.

In collaboration with H. Träuble, E. Sackmann, and W. Stoffel, the cooperative order ↔ disorder phase transition of the phospholipids could be readily demonstrated in membranes and cells (Fig. 2.2.1 and Refs. [7–9]).

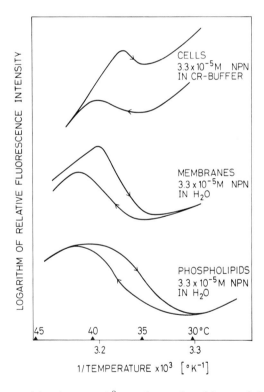

**Fig. 2.2.1:** Phase transition in *trans*-Δ⁹-octadecenoic acid-containing cells, membranes, and phospholipids of *E. coli* using N-phenyl-1-naphtylamine as a probe. See Ref. [8] for further details.

The fact that the temperature range of this transition was about the same in membranes and dispersions of isolated phospholipids suggested that proteins did not greatly perturb the interaction of the phospholipids. This surprising result can possibly be explained by the now well-known arrangement of proteins in the form of membrane-spanning α-helical segments. About 80% of the phospholipids in the membrane took part in the transition and, therefore, could be assumed to be arranged in a lipid bilayer (Fig. 2.2.2. and Ref. [10]).

The temperature range of the transition varied as a function of the hydrocarbon chain composition in a way expected from the behavior of synthetic model phospholipids. The bacteria could grow at a temperature far above the upper end of the phase transition or when up to 50% of the phospholipids were in ordered domains. Therefore, cell growth required that at least 50%

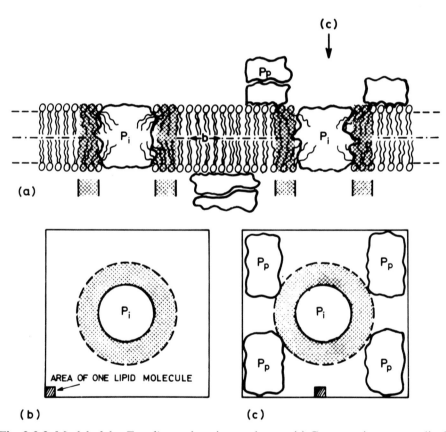

**Fig. 2.2.2:** Model of the *E. coli* cytoplasmic membrane. (a) Cross section perpendicular to the plane of the membrane with integral ($P_i$) and peripheral membrane proteins ($P_p$) in a bilayer of disturbed (crosshatched) and undisturbed phospholipids. (b) and (c) Top views of the membrane. See Ref. [10] for further details.

of the phospholipids were in a fluid state; ordered lipid domains were tolerated but not required. The implication of these studies was that the structure of membrane phospholipids had evolved in such a way as to guarantee a fluid state of the membrane at the growth temperature of the organism. Since the growth temperature is generally well above the order $\longleftrightarrow$ disorder transition, i. e. essentially all phospholipids are in the fluid state, the phase transition as such did not and still does not appear to have direct relevance for membrane function.

Cooling *E. coli* cells below the lipid phase transition strongly reduced the rate of protein-dependent functions of the membrane, such as the transport of β-galactosides by lactose permease (Fig. 2.2.3, Refs. [6–8]). The physical interpretation of the breaks in Arrhenius plots has remained controversial [11, 12]. By studying the anisotropy decay of pyrene-labeled lactose per-

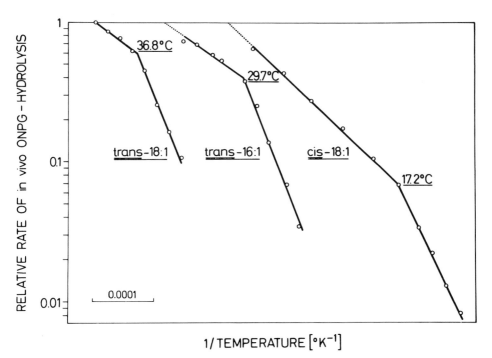

**Fig. 2.2.3:** Temperature dependence of o-nitrophenyl-β-D-galactoside (oNPG) hydrolysis in *E. coli* cells auxotrophic for unsaturated fatty acids. The cells were grown in the presence of *trans*-$\Delta^9$-octadecenoic acid (*trans*-18:1), *trans*-$\Delta^9$-hexadecenoic acid (*trans*-16:1) or *cis*-$\Delta^9$-octadecenoic acid (*cis*-18:1) as supplement. The rate of oNPG hydrolysis is a measure of the transport rate *via* lactose permease located in the cytoplasmic membrane. The temperature of characteristic changes in the slope in Arrhenius plots (log relative rate *versus* reciprocal absolute temperature) is indicated. Taken from Ref. [8].

mease, Dornmair and Jähnig [13] recently observed that a motion of about 50 nsec in the protein, most likely corresponding to the wobbling of a membrane-spanning α-helix, is reduced when the membrane is cooled below the phase transition. This experiment directly demonstrates that "solidification" of the lipid phase affects the motional properties of lactose permease, which are presumably part of the conformational change during sugar transport.

The transition points in Arrhenius plots of β-galactoside transport could be used to probe the lipid environment of the transport protein. In collaboration with F. F. Hill and I. Lamnek-Hirsch, it was found [14] that upon a change in the phospholipid hydrocarbon chain composition induced by a replacement in the medium-supplemented fatty acid, lactose permease was always exposed to a mixed lipid phase whether it was incorporated into the membrane before or during the change. Thus, on a time scale of 15–20 min, these experiments provided a strong argument for the randomization of the lipid phase by lateral diffusion. In fact, the lateral diffusion coefficient of phospholipids in *E. coli* membranes, $D = 3.25 \times 10^{-8}$ cm$^2$/sec, turned out to be similar to that in fluid model membranes of dipalmitoylphosphatidylcholine [9].

In summary, the experiments with the bacterial system provided one line of arguments for the now-established concept that in biological membranes phospholipids are arranged in a fluid lipid bilayer, which is both the barrier separating the inside and outside of a cell, and the matrix for the integration and attachment of proteins.

# References

[1]  K. Klein, R. Steinberg, B. Fiethen, P. Overath: Fatty acid degradation in *Escherichia coli*. An inducible system for the uptake of fatty acids and further characterization of *old* mutants. Eur. J. Biochem. 19 (1971) 442–450.

[2]  G. Pauli, P. Overath: *ato* operon: a highly inducible system for acetoacetate and butyrate degradation in *Escherichia coli*. Eur. J. Biochem. 29 (1972) 553–562.

[3]  T. van den Boom, J. E. Cronan, Jr.: Genetics and regulation of bacterial lipid metabolism. Ann. Rev. Microbiol. 43 (1989) 317–343.

[4]  H. U. Schairer, B. A. Haddock: β-Galactoside accumulation in a $Mg^{2+}$-, $Ca^{2+}$-activated ATPase deficient mutant of *E. coli*. Biochem. Biophys. Res. Commun. 48 (1972) 544–551.

[5]  B. A. Haddock, H. U. Schairer: Electron-transport chains of *Escherichia coli*. Reconstitution of respiration in a 5-aminolaevulinic acid-requiring mutant. Eur. J. Biochem. 35 (1973) 34–45.

[6]  H. U. Schairer, P. Overath: Lipids containing *trans*-unsaturated fatty acids change the temperature characteristic of thiomethylgalactoside accumulation in *Escherichia coli*. J. Mol. Biol. 44 (1969) 209–214.

[7]   P. Overath, H. U. Schairer, W. Stoffel: Correlation of *in vivo* and *in vitro* phase transitions of membrane lipids in *Escherichia coli*. Proc. Natl. Acad. Sci. USA 67 (1970) 606–612.
[8]   P. Overath, H. Träuble: Phase transitions in cells, membranes, and lipids of *Escherichia coli*. Detection by fluorescent probes, light scattering, and dilatometry. Biochemistry 12 (1973) 2625–2634.
[9]   E. Sackmann, H. Träuble, H. J. Galla, P. Overath: Lateral diffusion, protein mobility, and phase transitions in *Escherichia coli* membranes. A spin label study. Biochemistry 12 (1973) 5360–5369.
[10]  H. Träuble, P. Overath: The structure of *Escherichia coli* membranes studied by fluorescence measurements of lipid phase transitions. Biochim. Biophys. Acta 307 (1973) 491–512.
[11]  L. Thilo, H. Träuble, P. Overath: Mechanistic interpretation of the influence of lipid phase transitions on transport functions. Biochemistry 16 (1977) 1283–1290.
[12]  F. Jähnig, J. Bramhall: The origin of a break in Arrhenius plots of membrane processes. Biochim. Biophys. Acta 690 (1982) 310–313.
[13]  K. Dornmair, F. Jähnig: Internal dynamics of lactose permease. Proc. Natl. Acad. Sci. USA 86 (1989) 9827–9831.
[14]  P. Overath, F. F. Hill, I. Lamnek-Hirsch: Biogenesis of *E. coli* membrane: Evidence for randomization of lipid phase. Nature New Biol. 234 (1971) 264–267.

## Doctoral Theses

Georg Pauli (1972): Fettsäureabbau in *Escherichia coli*.
Klaus Klein (1973): Genetische und biochemische Charakterisierung der Flavoproteine (Acyl-CoA-Dehydrogenase und Elektronentransportierendes Flavoprotein) des Fettsäureabbaus in *Escherichia coli*.

## Postdoctoral Scientists

Ken Devor
Bruce Haddock
Hannelore Würz

# 2.3 Studies on Regulation of Gene Expression and on Membrane Biosynthesis in *Escherichia coli*

Ruth Ehring*

## 2.3.1 Summary

As will be summarized below, our group has pursued three main lines of research:
(1) Regulation of *E. coli* gene expression, especially as affected by bacterial insertion elements,
(2) biosynthesis and membrane insertion of *E. coli* lactose permease and
(3) protein/DNA interactions in the control of the *E. coli* galactose operon, in particular protein-induced changes of DNA conformation. Most of our work has been a collaborative effort in association with other groups of the Sonderforschungsbereich 74.

## 2.3.2 Regulation of Expression of the Galactose Operon *in vitro*

A major tool for our studies of insertion mutations (see below) was the DNA-directed system for cell-free synthesis of bacterial enzymes developed by Geoffrey Zubay and co-workers. Most of the insertion mutations characterized by Peter Starlinger and co-workers were located in the *E. coli* galactose operon. To define any effects exerted by such DNA elements on the expression of one or more of the three genes, we demonstrated first that synthesis of the three galactose enzymes was regulated by *gal* repressor and by cyclic AMP and responded as expected to mutations affecting regulatory units. Synthesis *in vitro* reflected the effects of controls known to be operative *in vivo*. It may be worth noting that this cell-free system represents a

---

* Institut für Genetik der Universität zu Köln

very sensitive assay system for the function of *gal* repressor that we have also utilized in more recent studies cited below. (Co-workers W. Wetekam and K. Staack, Refs. [1–3]).

## 2.3.3    Messenger RNA-Directed Enzyme Synthesis and Changes of RNA Conformation Affecting Translational Efficiency

Functional mRNA was found to exhibit a significantly longer half-life in the cell-free system than *in vivo*. This observation led us to develop one of the first systems suitable for mRNA-directed synthesis of bacterial proteins, which could be assayed by their catalytic activity. We could show reversible changes of *gal* mRNA conformation to differentially affect translational efficiencies. (Co-workers G. Schumacher and A. Krebs, Refs. [4, 5]).

## 2.3.4    Study of Polar Effects of Insertion Mutations

Mutants known to carry different insertion elements integrated into the *Escherichia coli* galactose operon had been characterized and were kindly provided by Drs. Starlinger, Saedler, Pfeifer, Besemer, and their colleagues. Our first experiments led to a new interpretation of the phenotype of mutations located in the control region of the operon (referred to as $0°$). Additionally, we could show that the polar effects exhibited by this class of IS-elements, as well as by those located within structural genes, depended on an activity requiring the *suA* gene product, which was later identified by others as the transcription termination factor Rho. (Co-workers W. Wetekam and K. Staack, Refs. [6, 7]).

## 2.3.5    Identification of an *IS4* Encoded Protein

One of the *E. coli* insertion elements studied intensively in Starlinger's group was *IS4*. It occurs with a low copy number and an unusually high extent of site-specificity. In collaboration with Paul Habermann of Starlinger's group and with Konrad Beyreuther, Klaus Trinks was able to show that an $M_r$

47 000 protein preferentially radiolabeled in plasmid-bearing *E. coli* mini-cells is the gene product encoded by the major open reading frame of the *IS4* DNA-element, which may be involved in the transposition process. Neither a functional test nor a specific antiserum were available at the time to detect this protein. The identification became possible by application of highly sensitive radiolabel amino acid sequence analysis performed on very small amounts of protein radiolabeled in plasmid-bearing *E. coli* minicells. (Co-worker Trinks in cooperation with Beyreuther and with research group of Starlinger, Refs. [8, 16]).

## 2.3.6 Identification of Proteins Predicted from DNA Sequences

The methods described in the preceding section, preferential expression of proteins in plasmid-bearing *E. coli* minicells or in DNA-directed cell-free protein synthesis, frequently in combination with radiolabeled amino acid sequences analysis, were also applied in a number of collaborative projects to identify and characterize several other gene products, which had been predicted from the analysis of their coding sequences. (Co-workers Trinks and Stochaj in cooperation with several research groups mostly within the Sonderforschungsbereich 74, Refs. [10 – 17]).

## 2.3.7 *E. coli* Lactose Permease is Synthesized Without Cleavable Signal Sequence

Lactose permease of *E. coli* was of particular interest among the gene products mentioned in the preceding paragraph, as it had served for many studies on mechanism and energetics of its transport function. Several studies were performed showing that the protein is synthesized without a cleavable N-terminal signal peptide and identifying Cys 148 as the cysteine residue specifically protected by substrate against chemical modification. Further studies concerned the accessibility of membrane bound lactose permease to proteolytic attack. (Co-worker Stochaj in collaboration with research groups of Beyreuther, Müller-Hill and Overath (Tübingen, FRG), Refs. [12–17].

## 2.3.8    Studies on Membrane Insertion of *E. coli* Lactose Permease

Subsequently, we have chosen lactose permease as an example of an integral membrane protein to study its biosynthesis and membrane insertion. Membrane attachment of the nascent protein was demonstrated to occur before translation of the protein is completed. To define more precisely regions important for insertion of the growing polypeptide chain into the lipid bilayer, we have used mutations specifically constructed to express incomplete *lacY*-related polypeptides. N-terminal segments of lactose permease were found to be capable of interacting with the membrane and to exhibit a certain degree of autonomy in this process.

   These studies have been extended recently to test combinations of incomplete *lacY*-related polypeptides synthesized simultaneously within the same cell for functional complementation. The requirements for such a successful interaction between artificially constructed subunits of an integral membrane protein are currently under investigation. (Co-workers U. Stochaj, U. Sonnewald, A. von Schaewen, C. (Heibach) Theres and W. Wrubel in cooperation with H.-J. Fritz and with members of U. Henning's group in Tübingen (FRG), Refs. [18–21]).

## 2.3.9    DNA/Protein Interactions in the Control of the Galactose Operon: RNA Polymerase and *GAL* Repressor Induce DNA-Bending

This study has been initiated jointly with Dr. Hans-Joachim Fritz (now Georg-August-Universität Göttingen), while he was at this institute. We have continued to cooperate in this work. The *E. coli* galactose genes are transcribed alternatively from one of two promoters, which are subject to superimposed positive and negative controls. Using oligonucleotide directed construction of structurally predetermined mutations, we are analyzing the contributions of the individual control elements. Of special interest is the role of the second operator, which was observed by Fritz and by others within the first structural gene of the operon.

   In the course of these studies, we found that both RNA polymerase and *gal* repressor bind to their respective recognition sites in the *gal* operon with DNA bending, as was previously known for the cyclic AMP receptor protein. Our present experiments concern possible effects on gene expression of such conformational changes introduced into DNA by the three proteins governing the expression of the galactose operon. (Co-workers: G. Kuhnke, C. (Heibach) Theres, S. Rosahl, A. Krause and W. Barth in cooperation with H.-J. Fritz and members of his research group, now at Göttingen University, Refs. [22–25]).

# 2.3.10 References

Many publications to which our group contributed reflect cooperation between different groups, mostly within the Sonderforschungsbereich 74. References are not given in chronological order but are roughly grouped according to topics.

Doctoral theses are listed below. Thirteen diploma theses (not listed separately) concern various aspects of the projects summarized above.

[1] W. Wetekam, K. Staack, R. Ehring: DNA-dependent *in vitro* synthesis of enzymes of the galactose operon of *Escherichia coli*. Mol. Gen. Genet. 112 (1971) 14–27.

[2] W. Wetekam, R. Ehring: Coordinate regulation of DNA-dependent cell-free synthesis of uridylyltransferase and galactokinase. FEBS Lett. 18 (1971) 271–273.

[3] W. Wetekam: Identification of template strand in heteroduplex DNA directing cell-free enzyme synthesis. Mol. Gen. Genet. 118 (1972) 57–60.

[4] G. Schumacher, R. Ehring: RNA directed cell-free synthesis of the galactose enzymes of *Escherichia coli*. Mol. Gen. Genet. 124 (1973) 329–344.

[5] G. Schumacher, R. Ehring: Effect of different conformations of galactose messenger RNA on gene expression and messenger half-life *in vitro*. Mol. Gen. Genet. 136 (1975) 41–54.

[6] W. Wedekam, K. Staack, R. Ehring: Relief of polarity in DNA-dependent cell-free synthesis of enzymes of the galactose operon of *Escherichia coli*. Mol. Gen. Genet. 116 (1972) 258–276.

[7] W. Wetekam, R. Ehring: A role for the product of gene *suA* in restoration of polarity *in vitro*. Mol. Gen. Genet. 124 (1973) 345–358.

[8] K. Trinks, P. Habermann, K. Beyreuther, P. Starlinger, R. Ehring: An *IS4*-encoded protein is synthesized in minicells. Mol. Gen. Genet. 182 (1981) 183–188.

[9] G. Pauli, R. Ehring, P. Overath: Fatty acid degradation in *Escherichia coli*: requirement of cyclic AMP and cyclic AMP receptor protein for enzyme synthesis. J. Bact. 117 (1974) 1178–1183.

[10] G. Schröder, W. Klipp, A. Hillebrand, R. Ehring, C. Koncz, J. Schröder: The conserved part of the T-region in Ti plasmids expresses four proteins in bacteria. EMBO J. 2 (1983) 403–409.

[11] J. Schröder, W. Klipp, A. Hillebrand, R. Ehring, C. Koncz, G. Schröder: Expression of the T-region of Octopine plasmid pTi Ach5 into protein in bacterial systems. In: A. Pühler (ed.): *Molecular genetics of the bacteria plant interaction*. Springer-Verlag, Berlin-Heidelberg 1983, pp. 259–267.

[12] R. Ehring, K. Beyreuther, J. K. Wright, P. Overath: *In vitro* and *in vivo* products of lactose permease gene are identical. Nature 283 (1980) 537–540.

[13] K. Beyreuther, R. Ehring, P. Overath, J. K. Wright: Microsequence analysis of lactose permease of *Escherichia coli*. In: C. Birr (ed.): *Methods in peptide and protein sequence analysis:* Elsevier/North-Holland, Amsterdam 1980, pp. 199–212.

[14] K. Beyreuther, B. Bieseler, R. Ehring, H.-W. Grießer, M. Mieschendahl, B. Müller-Hill, I. Triesch: Investigation of structure and function of lactose permease of *Escherichia coli*. Biochem. Soc. transactions 8 (1980) 675–676.

[15] K. Beyreuther, B. Bieseler, R. Ehring, B. Müller-Hill: Identification of internal residues of lactose permease of *Escherichia coli* by radiolabel sequencing of peptide mixtures. In: M. Elsinga (ed.): *Methods in protein sequence analysis:* Humana Press, Inc. Clifton, N. Y. 1980, pp. 139–148.

[16] K. Beyreuther, K. Stüber, B. Bieseler, J. Bovens, R. Dildrop, T. Geske, I. Triesch, K. Trinks, S. Zaiss, R. Ehring: From genes to proteins: genotypic and phenotypic analysis of DNA sequences by protein sequencing. In: U. Jensen, D. E. Fairbrothers (eds.): *Proteins and nucleic acids in plant systematics.* Springer-Verlag, Berlin-Heidelberg 1983, pp. 85–104.

[17] U. Stochaj, B. Bieseler, R. Ehring: Limited proteolysis of lactose permease from *Escherichia coli*. Eur. J. Biochem. 158 (1986) 423–428.

[18] U. Stochaj, R. Ehring: The N-terminal region of *Escherichia coli* lactose permease mediates membrane contact of the nascent polypeptide chain. Eur. J. Biochem. 163 (1987) 653–658.

[19] U. Stochaj, H.-J. Fritz, C. Heibach, M. Markgraf, A. von Schaewen, U. Sonnewald, R. Ehring: Truncated forms of *Escherichia coli* lactose permease: models for study of biosynthesis and membrane insertion. J. Bacteriol. 170 (1988) 2639–2645.

[20] S. MacIntyre, M.-L. Eschbach, H. Schwarz, R. Ehring: Topological analysis of the amino-terminal region of lactose permease using the *Escherichia coli* outer membrane protein, OmpA, as a marker. FEBS Lett. 247 (1989) 396–400.

[21] W. Wrubel, U. Stochaj, U. Sonnewald, C. Theres, R. Ehring: Reconstitution of an active lactose carrier *in vivo* by simultaneous synthesis of two complementary protein fragments. Journ. Bacteriol. 172 (1990) 5374–5381.

[22] H.-J. Fritz, H. Bicknäse, B. Gleumes, C. Heibach, S. Rosahl, R. Ehring: Characterization of two mutations in the *Escherichia coli galE* gene inactivating the second galactose operator and comparative studies of repressor binding. EMBO J. 2 (1983) 2129–2135.

[23] G. Kuhnke, A. Krause, C. Heibach, U. Gieske, H.-J. Fritz, R. Ehring: The upstream operator of the *Escherichia coli* galactose operon is sufficient for repression of transcription initiated at the cyclic AMP-stimulated promoter. EMBO J. 5 (1986) 167–173.

[24] G. Kuhnke, H.-J. Fritz, R. Ehring: Unusual properties of promoter-up mutations in the *Escherichia coli* galactose operon and evidence suggesting RNA polymerase-induced DNA bending. EMBO J. 6 (1987) 507–513.

[25] G. Kuhnke, C. Theres, H.-J. Fritz, R. Ehring: RNA polymerase and *gal* repressor bind simultaneously and with DNA bending to the control region of the *Escherichia coli* galactose operon. EMBO J. 8 (1989) 1247–1255.

*Ruth Ehring*

## Doctoral Theses

Waldemar Wetekam (1973): Studien zur Polaritätsausprägung in einem zellfreien System für die Synthese der Enzyme des Galaktose Operons.

Günter Schumacher (1976): Untersuchungen zur Genausprägung in einem zellfreien RNA-abhängigen System für die Synthese der Galactose Enzyme von *E. coli.*

Klaus Trinks (1981): Untersuchungen zur Proteinsynthese *in vitro* und in Minizellen von *Escherichia coli:* Identifizierung eines IS4-codierten Proteins.

Ursula Stochaj (1985): Biosynthese und Membranassoziation der Laktose Permease von *Escherichia coli.*

Günter Kuhnke (1988): Protein/DNA-Wechselwirkungen in der Kontrollregion des *E. coli* Galaktose Operons.

## Present Affiliation of Former Co-workers

Dr. Waldemar Wetekam, Forschungslaboratorium der Hoechst A.G., Frankfurt a. Main 80, FRG.

Dr. Günter Schumacher, Forschungslaboratorium der Fa. Boehringer Mannheim, Penzberg, FRG.

Dr. Anita Krebs, Queen's University, Kingston, Canada K7L 3N6.

Dr. Klaus Trinks, Pflanzenschutzforschung Hoechst A.G., Frankfurt a. Main 80, FRG.

Dr. Ursula Stochaj, Princeton University, Department of Biology, Princeton, NJ. 08544 USA.

## Visiting Scientists

Dr. Ursula Stochaj, Institut für Anatomie und Zellbiologie, Universität Marburg, FRG.

Dipl.-Biol. Theodor Hanck, Institut für Molekulare Genetik, Universität Göttingen, FRG.

# 2.4     The Beginnings of Protein Engineering

Bernd Gutte*

## 2.4.1     Summary

The work summarized in this chapter was performed by my group between 1972 and 1977, when I was a member of the Sonderforschungsbereich 74. It dealt with structure-function studies of ribonuclease A and insulin and is considered an early example of protein engineering [1].

## 2.4.2     Truncated, Biologically Active Ribonuclease A

### 2.4.2.1     Introduction

We were interested in the question of whether deletion analogues of RNase A (ribonuclease A, 124 amino acid residues) retained the ability to adopt native-like structures. The deletions which comprised the amino terminus and four peripheral loops were determined by inspection of a PCK model of the x-ray structure of RNase [2].

### 2.4.2.2     Results

Several 70- and 63-residue RNase A analogues [3–5] were synthesized by the solid phase method [6–8]. The sequences of one of the 63-residue analogues [4] and of the natural enzyme are given in Fig. 2.4.1.

---

*Institut für Genetik der Universität zu Köln
(at present: Biochemisches Institut der Universität Zürich)

```
26                         35   41
C-N-Q-M-M-K-S-R-N-L-A-K-P-V-N-T-F-V-H-E-S
                5973                     84
L-A-D-V-Q-A-V-C-S-Y-Q-S-Y-S-T-M-S-I-T-D-C
 100                   110 117          124
G-T-Q-A-N-K-H-I-I-V-A-C-G-P-V-H-F-D-A-S-V
```

```
1
K-E-T-A-A-A-K-F-E-R-Q-H-M-D-S-S-T-S-A-A-S
22
S-S-N-Y-C-N-Q-M-M-K-S-R-N-L-T-K-D-R-C-K-P
43
V-N-T-F-V-H-E-S-L-A-D-V-Q-A-V-C-S-Q-K-N-V
64
A-C-K-N-G-Q-T-N-C-Y-Q-S-Y-S-T-M-S-I-T-D-C
85
R-E-T-G-S-S-K-Y-P-N-C-A-Y-K-T-T-Q-A-N-K-H
106                                   124
I-I-V-A-C-E-G-N-P-Y-V-P-V-H-F-D-A-S-V
```

**Fig. 2.4.1:** Amino acid sequences of the enzymatically active 63-residue deletion analogue of RNase A (top) and the natural enzyme (bottom).

Figure 2.4.2 shows the 63-residue analogue surrounded by the pieces that were omitted. Only one of the deletions (residues 1 to 25) contained amino acids thought to be essential for substrate binding and catalysis [9]. The synthetic 63-residue RNase A analogue was found to have the following properties [4]:

– It was bound by an affinity column specific for RNase A.
– It cleaved RNA with RNase A-like specifity and had approximately 10 % of the activity of the natural 124-residue enzyme.
– It cross-reacted strongly with anti-RNase A antibodies (G. W. Welling and B. Gutte, unpublished results).

## 2.4.2.3 Conclusions

From the results obtained, the following conclusions can be drawn:

– The synthetic 63-residue "fusion protein" (Fig. 2.4.2, center) and the corresponding portion of the natural enzyme seemed to be very similar in

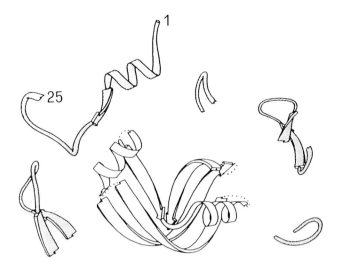

**Fig. 2.4.2:** 63-residue RNase A (bottom center) is surrounded by the pieces that were omitted. Arrows are β-strands and spirals are α-helices. (Reproduced with permission from Trends Biochem. Sci. 14 (1989) 305.)

structure because the 63-residue analogue had RNase A activity and cross-reacted with anti-RNase A antibodies.

– The deletions included 27 residues that were invariant in 30 mammalian ribonucleases [10, 11].

– Replacement of lysine 41 and lack of histidine 12, two residues considered to be essential for transition state stabilization and catalysis [9], did not abolish the enzymatic activity of the 63-residue RNase A analogue.

Recently, another example of a biologically active protein with an incomplete ligand binding site was reported [12]. Single heavy chain variable domains, prepared by genetic engineering, had 10 % of the lysozyme affinity of complete mouse anti-lysozyme antibodies (the dissociation constants were 20 nM and 2nM, respectively).

*Bernd Gutte*

## 2.4.3 Combination of the A and B Chain of Insulin on an Anti-Insulin Antibody Column

### 2.4.3.1 Introduction

The combination of insulin A and B chain in solution (for review, see Ref. [13]) yields a mixture of disulfide isomers of insulin from which the authentic, biologically active hormone is difficult to isolate. Therefore, we have developed a method that allows the exclusive formation of the three disulfide bonds of insulin [14].

### 2.4.3.2 Results

It was known that the B chain of insulin crossreacts with anti-insulin antibodies, whereas the A chain does not [15]. If the B chain were held by the antibodies in a native-like conformation, fewer by-products would be expected to form during the reaction with the A chain. The combination experiment was carried out as follows [14]: S-sulfonate B chain ($B(S-SO_3^-)_2$) was bound to Sepharose 4B-linked anti-insulin antibodies and then reacted at pH 8.3 with a 20-fold molar excess of reduced A chain ($A(SH)_4$). The products were excess A chain in the oxidized form ($A(S-S)_2$), unreacted $B(S-SO_3^-)_2$, and insulin (yield, 24 %). The identity of the resynthesized hormone was demonstrated by protein-chemical, physico-chemical, and biological assays.

### 2.4.3.3 Conclusions

The chain combination method described above provides pure insulin in a single step and therefore is superior to multistep procedures that use reversible cross-bridging between the A and B chain to assist in the formation of the disulfide pattern of insulin [16, 17]. Experiments on a preparative scale have not yet been performed. The suitability of shortened proinsulins [18] as starting material for an economical synthesis of insulin still has to be explored.

*Acknowledgements*

I thank my collaborators Ronald E. Reid, Mechthild Regenass-Klotz, Willi Kullmann, Klaus-Rüdiger Bärwald, Johannes Döhmer and Marlies Hoffmann.

# 2.4.4   References

[1] J. S. Richardson, D. C. Richardson: The *de novo* design of protein structures. Trends Biochem. Sci. 14 (1989) 304–309.
[2] H. W. Wyckoff, D. Tsernoglou, A. W. Hanson, J. R. Knox, B. Lee, F. M. Richards: The three-dimensional structure of ribonuclease-S. J. Biol. Chem. 245 (1970) 305–328.
[3] B. Gutte: A synthetic 70-amino acid residue analog of ribonuclease S-protein with enzymic activity. J. Biol. Chem. 250 (1975) 889–904.
[4] B. Gutte: Study of RNase A mechanism and folding by means of synthetic 63-residue analogs. J. Biol. Chem. 252 (1977) 663–670.
[5] B. Gutte: Synthetic 63-residue RNase A analogs. J. Biol. Chem. 253 (1978) 3837–3842.
[6] R. B. Merrifield: Solid phase peptide synthesis. I. The synthesis of a tetrapeptide. J. Am. Chem. Soc. 85 (1963) 2149–2154.
[7] B. W. Erickson, R. B. Merrifield: Solid-phase peptide synthesis. In: H. Neurath, R. L. Hill: The Proteins, 3rd edn., Volume II. Academic Press, New York 1976, pp. 255–527.
[8] R. E. Reid: Solid phase peptide synthesis. J. Org. Chem. 41 (1976) 1027–1031.
[9] F. M. Richards, H. W. Wyckoff: Bovine pancreatic ribonuclease. In: P. D. Boyer: The Enzymes, 3rd edn., Volume IV. Academic Press, New York 1971, pp. 647–806.
[10] F. Russchen, G. de Vrieze, W. Gaastra, J. J. Beintema: Studies on the covalent structure of eland pancreatic ribonuclease. Biochim. Biophys. Acta 427 (1976) 719–726.
[11] B. Gutte: Synthesis and enzymic properties of a 63-residue analogue of ribonuclease A. Biochem. Soc. Trans. 3 (1975) 897–899.
[12] E. S. Ward, D. Güssow, A. D. Griffiths, P. T. Jones, G. Winter: Binding activities of a repertoire of single immunoglobulin variable domains secreted from *Escherichia coli*. Nature 341 (1989) 544–546.
[13] K. Lübke, H. Klostermeyer: Synthese des Insulins – Anfänge und Fortschritte. Adv. Enzymol. Relat. Areas Mol. Biol. 33 (1970) 445–525.
[14] M. Klotz, B. Gutte: Combination of insulin chains on an anti-insulin antibody Sepharose column. Nature 262 (1976) 791–793.
[15] Y. Yagi, P. Maier, D. Pressman: Antibodies against the component polypeptide chains of bovine insulin. Science 147 (1965) 617–619.
[16] R. Geiger, R. Obermeier: Insulin synthesis from natural chains by means of reversible bridging compounds. Biochem. Biophys. Res. Commun. 55 (1973) 60–66.
[17] D. Brandenburg, W. Schermutzki, H. Zahn: $N^{\alpha A1}$-$N^{\epsilon B29}$-crosslinked diaminosuberoylinsulin, a potential intermediate for the chemical synthesis of insulin. Hoppe-Seyler's Z. Physiol. Chem. 354 (1973) 1521–1524.
[18] W. Kullmann, B. Gutte: A shortened synthetic proinsulin. Biochem. Soc. Trans. 3 (1975) 899–902.

*Bernd Gutte*

## Doctoral Students

Mechthild Regenass-Klotz, Biozentrum der Universität Basel, Switzerland; private life.
Willi Kullmann, Max-Planck-Institut für Biophysikalische Chemie, Göttingen, FRG;
    Institut für Zellbiochemie des Universitätskrankenhauses Hamburg-Eppendorf,
    Hamburg, FRG. Habilitation 1986 at the Universität Göttingen.

## Postdoctoral Scientists

Dr. R. E. Reid, Associate Professor at the Faculty of Pharmacy, University of Manitoba,
    Winnipeg, Canada.

## 2.5 Membrane Lipids, Lipoproteins and Membrane Structures
## The Targets of 15 Years of Biochemical Research

Wilhelm Stoffel*

## 2.5.1 Preface

This is the final report on the results of my work achieved within the Sonderforschungsbereich 74 "Molekularbiologie der Zelle". I would like to begin by expressing my sincere gratitude to the Deutsche Forschungsgemeinschaft. Its financial support made it possible for me and the young scientists who joined my group over the years to do research in a relatively unrestricted way and to effectively utilize our biochemical know-how and broaden our methodology in a problem solving way. We were able to expand outside the established field of lipid biochemistry into the field of protein chemistry and to establish successfully the methods of molecular biology.

I would particularly like to acknowledge the efforts of my reviewers, who evaluated the achievements of my laboratory favourably, but critically, over the years, and who, in addition, on the occasion of the several site visits, supported me with stimulating and competent discussions.

This is particularly the case for the rich scientific exchange and personal interaction between my group and others within the Sonderforschungsbereich. I would like to repeat my deep appreciation of the skilled collaboration with the permanent technical staff and the young scientists, mostly graduate students (Diploma and PH. D. students), who joined me over the years. Their contributions were acknowledged in the "Scientific Reports" (Arbeitsberichte) and the volumes "Molecular Biology in Cologne". Their contributions are documented appropriately in the respective publications.

---
* Institut für Biochemie, Medizinische Fakultät der Universität zu Köln

## 2.5.2 Summary of the Scientific Results of the Projects Supported within the Sonderforschungsbereich "Molekularbiologie der Zelle"

The results of my projects proposed for each of the three years periods are reported chronologically in the respective "Scientific Reports". In this final report, I would like to avoid their presentation as "piece meal" but try to outline how the problems evolved and how we solved them and what the experimental answers were.

Therefore, I limit and reduce this report to the answers to the following questions.

## 2.5.3 "What was Discovered in the Fields Supported by the Sonderforschungsbereich?"

### 2.5.3.1 Metabolism of Membrane Lipids

2.5.3.1.1 We Unravelled the *Biosynthesis* of Long Chain Sphingosine Bases[8–12]

Sphingosine bases are the key components of cellular complex lipids. Sphingolipids are, besides phospholipids and cholesterol, the main constituents of plasma membranes, particularly in neurons, astrocytes and oligodendroglia.

Figure 2.5.1 A presents the stereochemistry of the sphingosine bases (2S,3R-2-amino-1,3-dihydroxyoctadecane) and Fig. 2.5.1 B outlines the biosynthetic reaction sequence. First palmitoyl-CoA and serine are condensed by the pyridoxal phosphate dependent *3-keto-sphinganine-synthase,* a membrane bound enzyme of the endoplasmic reticulum membrane. Second the NADPH dependent reduction of 3-ketosphinganine to dihydroxysphingosine (=sphinganine) occurs.

The latter never occurs free in the cell but is immediately acylated by long chain acyl-CoA derivatives to ceramide. This acylation is catalysed again by a membrane bound acyltransferase. The saturated ceramide is the substrate for the dehydrogenase, yielding the predominant desaturated ceramide, the precursor for sphingomyelin, ceramide galactoside, sulfatide, ceramide polyhexoside, and ganglioside biosynthesis.

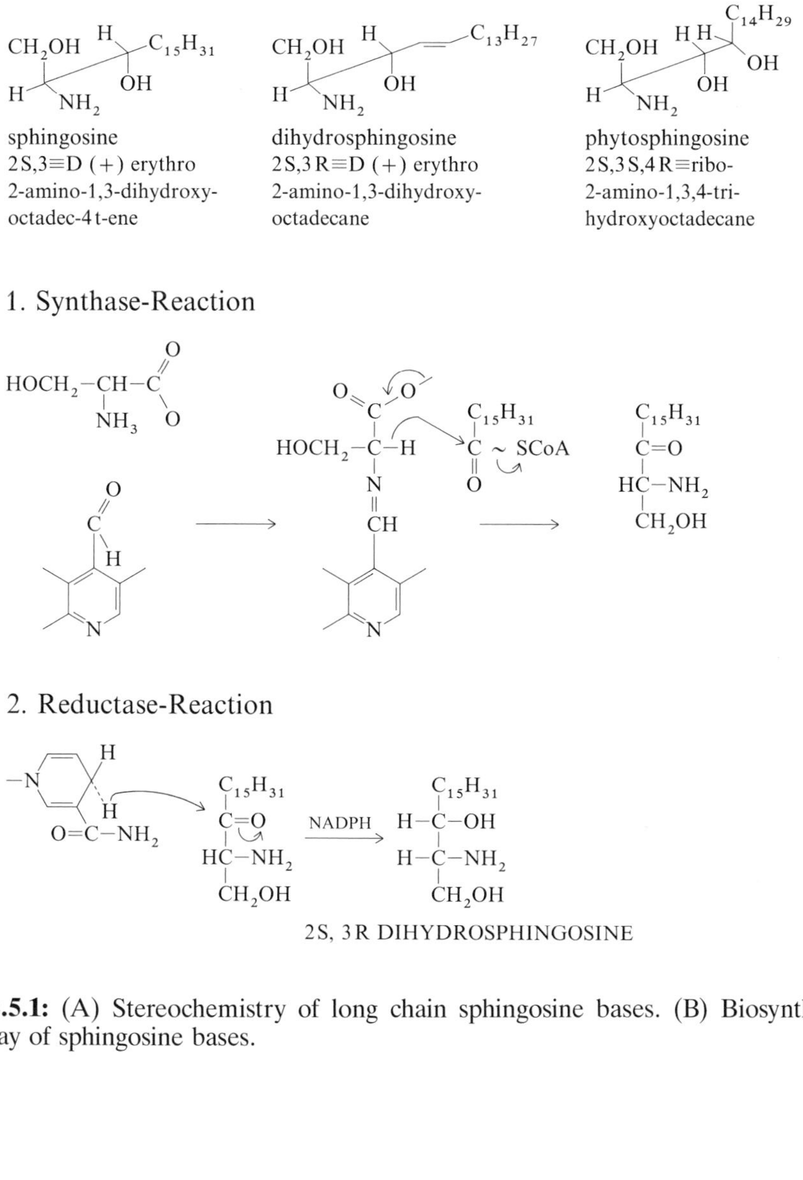

sphingosine
2S,3≡D (+) erythro
2-amino-1,3-dihydroxy-
octadec-4t-ene

dihydrosphingosine
2S,3R≡D (+) erythro
2-amino-1,3-dihydroxy-
octadecane

phytosphingosine
2S,3S,4R≡ribo-
2-amino-1,3,4-tri-
hydroxyoctadecane

## 1. Synthase-Reaction

## 2. Reductase-Reaction

2S, 3R DIHYDROSPHINGOSINE

**Fig. 2.5.1:** (A) Stereochemistry of long chain sphingosine bases. (B) Biosynthetic pathway of sphingosine bases.

### 2.5.3.1.2 We Discovered the Pathway of the *Degradation* of the Long Chain Sphingosine Bases

The pathway of the degradation of sphingosine bases is outlined in Fig. 2.5.2.

A soluble cytosolic *sphinganine-1-kinase* phosphorylates the primary alcohol group of the long chain bases. Sphingosine-1-phosphate is then cleaved in an aldol-like reaction catalyzed by the endoplasmic reticulum

### 1. Kinase-Reaction

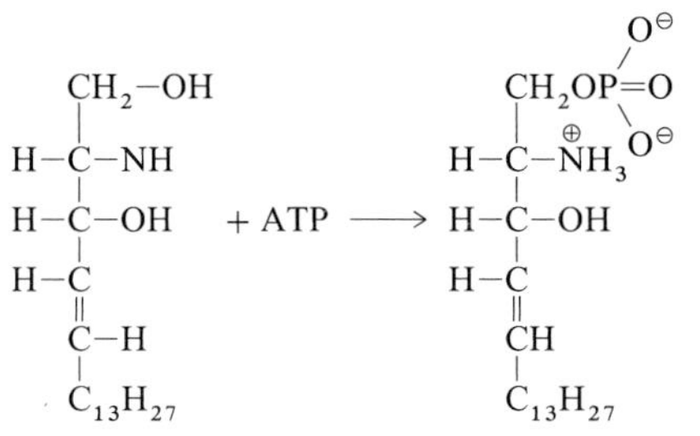

$$
\begin{array}{l}
CH_2-OH \\
H-C-NH \\
H-C-OH \quad + ATP \longrightarrow \\
H-C \\
\quad \| \\
\quad C-H \\
\quad C_{13}H_{27}
\end{array}
\qquad
\begin{array}{l}
CH_2OP\!=\!O \\
H-C-\overset{\oplus}{N}H_3 \\
H-C-OH \\
H-C \\
\quad \| \\
\quad CH \\
\quad C_{13}H_{27}
\end{array}
$$

### 2. Aldolase-Reaction

$$
\begin{array}{l}
CH_2-O-PO_3^{2-} \\
H-C-\overset{\oplus}{N}H_3 \\
H-C-OH \\
H-C \\
\quad \| \\
\quad CH \\
\quad C_{13}H_{27}
\end{array}
\qquad \longrightarrow \qquad
\begin{array}{l}
CH_2-O-PO_3^{2-} \\
H-C-N\!=\!CH- \\
H-C-OH \\
H-C \\
\quad \| \\
\quad CH \\
\quad C_{13}H_{27}
\end{array}
$$

$$
Enz\text{-}S \sim
\begin{array}{l}
\overset{H}{C}-OH \\
C \\
\| \\
C-H \\
C_{13}H_{27}
\end{array}
\quad + \quad
\begin{array}{l}
CH_2-N\!=\!CH- \\
CH_2 \\
O-PO^{2-}
\end{array}
$$

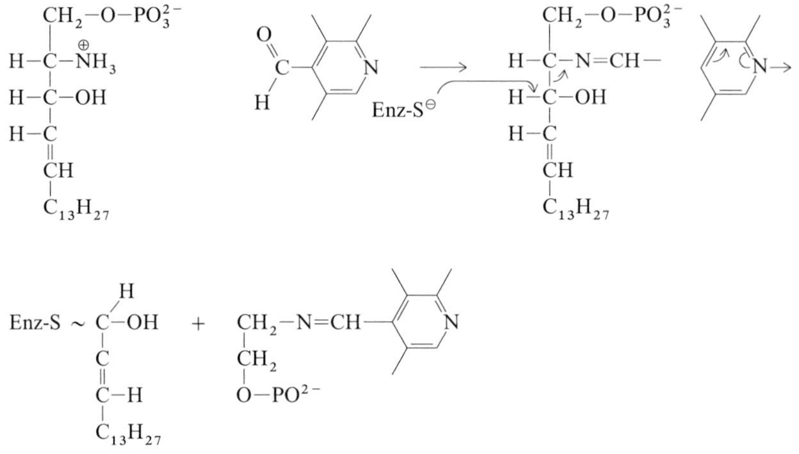

**Fig. 2.5.2:** Degradation pathway of sphingosine bases.

membrane bound aldolase to *phosphorylethanolamine* and either *palmitaldehyde* (dehydrosphingosine), *2-hexadecenal* (sphingosine) or *2-hydroxy palmitaldehyde* (phytosphingosine).

### 2.5.3.1.3  We Established Metabolic Links of Sphingosine Degradation Products to Phospholipid Biosynthesis

The long chain aldehyde products in the aldolase reaction can be oxidized to palmitic acid and re-utilized for phospholipid synthesis, degraded by β-oxidation or reduced to hexadecanol by alcohol dehydrogenase.

Hexadecanol is the substrate in the biosynthesis of alkyl and α-*cis alkenyl ether phospholipids* (plasmalogens). Isotope studies revealed the very efficient incorporations of the phosphoryl ethanol fragment released from sphingosine bases in the aldolase cleavage reaction.

## 2.5.3.2 Biophysical Studies on Membrane Lipids and Synthetic (Artificial) Bilayers[4,13-22,31]

### 2.5.3.2.1 We designed the Chemical Syntheses of Fluorescent and [13]C-labelled Fatty Acids of Variable Chain Lengths

– Using a newly developed acylation procedure we elaborated a new synthetic pathway for *fluorescent ω-labelled fatty acids* of different chain lengths and their incorporation into phospho- and sphingolipids.
– Following procedures for synthesis elaborated previously in my laboratory, *[13]C-labelled fatty acids*, which were enriched in specific positions along the alkane chain, were synthesized. Phospho- and sphingolipids were then acylated with these labelled fatty acids.

Also the polar head groups of phosphatidylcholine and sphingomyelin were enriched with [13]$CH_3$-groups following a procedure previously elaborated in my group. These [13]C-enriched lipid species made the experiments described below feasible, since studies would not be possible by measuring only the natural abundance of [13]C.

### 2.5.3.2.2 Fluorescence- and [13]C-NMR Spectroscopic Studies of Artificial Lipid Bilayers

Using a fluorescent labelled group at the ω-position and the [13]C-enriched isotope in different positions of the fatty acyl chains proved to be very sensitive indicators for probing the mobility of the lipid bilayer along the carbon chain between the carboxyl-group and the terminal $CH_3$ group.

The bulky anthryl group had limitations as a probe because of its perturbation contrary to the [13]C-isotope. $T_1$-relaxation time measurements gave precise information about the sequential mobility of the alkane chains within the lipid bilayer.

Our studies made essential contributions to our present view of the fluidity of biological membranes and the influence of the nature of the fatty acyl chains, the polar head groups of complex lipids, and the cholesterol content in the membranes. Studies on lipid-protein complexes with β-hydroxybutyrate dehydrogenase and glycophorin in [13]C-labelled phospholipid bilayers led to the *concept of the lipid halo around integral membrane proteins*. These data were also derived from $T_1$-relaxation time analyses of the reconstituted systems.

$^{13}$C-NMR-spectroscopy of human serum high density lipoproteins (HDL) enriched by a newly developed lipid exchange method with $^{13}$C-labelled phospholipids and 26-$^{13}$C-cholesterol, together with the structural studies described below, resulted in our present concept of the general molecular architecture of these lipid transporting particles. *A monomolecular phospholipid-cholesterol layer together with the particles' specific apolipoproteins (apo AI and AII), folded into amphipathic helices around the hydrophobic core, harbouring cholesterol esters and triglycerides.*

Our model presents the adequate molecular basis for all known functions of the HDL particle: the apo AI supported trans-esterification of cholesterol by the LCAT (lecithin-cholesterol-acyl transferase), the reversed cholesterol transport, and the receptor mediated recognition of the HDL particle required for its metabolism.

## 2.5.3.3   High-Density Lipoproteins[3,6,23–25,28,32]

### 2.5.3.3.1   We Performed Structural Studies on the Lipid Apoprotein Interactions and apo AI – apo AII Interactions

The localization and nearest neighbor relationships of the surface components of the HDL particle were analyzed using radioactive azido fatty acids introduced into phospholipids. The *acido group is photosensitive* and leads to a crosslinking reaction with the nearest neighbor. The position of the nitrene within the fatty acyl chain acts like a ruler, permitting the determination of the depth of the penetration of the amphipathic helices of the apoproteins.

Because the molecular properties of the crosslinking reagents determine the "radius of the action" of these probes, *bifunctional crosslinking reagents,* e. g. $^{14}$C-labelled dimethylsuberimidate and dithio-bis-butyrimidate or tartrate-diazide allowed us to determine the nearest neighboring domains of the surface-oriented apolipoproteins by protein-chemical analyses of the crosslinked proteolytic cleavage products. As a final result of these tedious studies *a map of the HDL surface* could be derived.

This principle of the construction of the HDL particle was also verified in other serum lipoprotein particles, e. g. LDL.

2.5.3.3.2 We Made Cell Biological Studies on the Synthesis, Translocation, Secretion and Processing of apo AI and apo AII of Human HDL

We discovered the processing and secretion pathway of apo AI and apo AII. Our structural studies were accompanied by the analysis of the *biogenesis and secretion of the HDL apolipoproteins.* During a short sabbatical in the laboratory of G. Blobel, Rockefeller University, New York, I succeeded in demonstrating that both apo AI and apo AII (as shown later in my own laboratory) were synthesized as prepro-apolipoproteins. The presequence was shown to be lost during cotranslational translocation. The pro apo AI was fully secreted and the pro apo AII partially secreted before the prosequences, a hexapeptide in the case of pro AI and a pentapeptide in the case of pro AII, were processed in the serum by a specific thiol protease. The amino acid sequences of the pre- and pro-sequences were determined by radio-micro sequencing and later confirmed by our own cloning techniques and that of other groups. The proposed biosynthetic pathway, derived from studies *in vitro*, was also confirmed by studies in tissue culture using Hep G2 cells.

2.5.3.3.3 We Investigated the Molecular Biology of apo AI and apo AII

*Full length apo AI and apo AII cDNA clones were isolated* from a human liver cDNA library which we constructed from human liver poly(A)$^+$ RNA using labelled oligonucleotides derived from the apo AI and AII structure as probes. *Their genes were isolated from genomic libraries.*

When the Sonderforschungsbereich 74 expired, we had reached the stage in which we studied and succeeded in the transient expression of the apo AI and AII genes in *Xenopus laevis* oocytes and in CHO cells. With the latter system the regulation of the expression is being studied particularly with regard to the influence of cholesterol, estradiol and insulin. We used the site directed mutagenesis to completely exchange the amino acid sequences which signal the pre- and prosequence of apo AI.

These studies convincingly demonstrated that the prosequence is neither required for the translocation nor the secretion of apo AI. Its function remains unclear. The size and conformation of the signal sequence however, is of decisive importance for the secretion and processing.

## 2.5.3.4 Molecular Neurobiology[1,2,7,26,27,30]

### 2.5.3.4.1 We Conducted Structural Studies on the Myelin Membrane of Central Nervous System (CNS) Myelogenesis

*We unraveled the primary structure of the integral myelin membrane protein Proteolipid Protein (PLP).* The lipid and protein constituents of the myelin membrane of CNS are synthesized by the oligodendroglia during a short myelinization period (day 10 to 28 after birth in rodents). This myelogenesis follows a well tuned regulation, programmed in time and space.

The lipid-rich 50 Å thick myelin membranes from spirally stacked insulating multilayer systems consisting of myelin-specific complex lipids, e. g. cerebroside, sulfatides in addition to phospholipids (including plasmologens), sphingomyelin and the high (40 mol%) cholesterol. The basic myelin protein (MBP) is attached as a peripheral membrane protein to the apposing cytosolic surfaces of the oligodendrocyte plasma membrane processes which induces a compaction of the cytosolic cleft by ionic interactions with the acid polar head groups of the lipid bilayer.

**A**

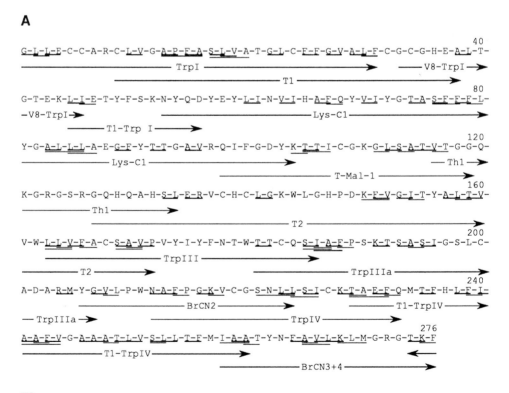

**B**

```
1        -         +
G-L-L-E-C-C-A-R-
```

**27 Aminosäuren; 7,5 Ganghöhen; 4,05 nm    27 aa.; 7.5 pitches; 4.05 nm**
```
    10              20              30
C-L-V-G-A-P-F-A-S-L-V-A-T-G-L-C-F-F-G-V-A-L-F-C-G-C-G-
```

```
+  -     40     - +      -    50  +       -   -
H-E-A-L-T-G-T-E-K-L-I-E-T-Y-F-S-K-N-Y-Q-D-Y-E-
```

**29 Aminosäuren; 8 Ganghöhen; 4,35 nm    29 aa.; 8 pitches; 4.35 nm**
```
   60        (+)        70              80
Y-L-I-N-V-I-H-A-F-Q-Y-V-I-Y-G-T-A-S-F-F-F-L-Y-G-A-L-L-L-A-
```

```
-    90              +    100   -    +            +              120+
E-G-F-Y-T-T-G-A-V-R-Q-I-F-G-D-Y-K-T-T-I-C-G-K-G-L-S-A-T-V-T-G-G-Q-K-
```
```
    +      +      (+)130(+)     -  +    (+)140    +       (+)  -  +
G-R-G-S-R-G-Q-H-Q-A-H-S-L-E-R-V-C-H-C-L-G-K-W-L-G-H-P-D-K-
```

**40 Aminosäuren; 11 Ganghöhen; 6,00 nm    40 aa.; 11 pitches; 6.00 nm**
```
            160             170
F-V-G-I-T-Y-A-L-T-V-V-W-L-L-V-F-A-C-S-A-V-P-V-
```
```
            180             190
Y-I-Y-F-N-T-W-T-T-C-Q-S-I-A-F-P-S-
```

```
+               200  -    +
K-T-S-A-S-I-G-S-L-C-A-D-A-R-
```

**12 Aminosäuren; 3 Ganghöhen; 1,65 nm    12 aa.; 3 pitches; 1.65 nm**
```
        210
M-Y-G-V-L-P-W-N-A-F-P-G-
```

```
+  220              +  230-           (+)
K-V-C-G-S-N-L-L-S-I-C-K-T-A-E-F-Q-M-T-F-H-
```

**30 Aminosäuren; 8,3 Ganghöhen; 4,50 nm    30 aa.; 8.3 pitches; 4.50 nm**
```
   240          250              260
L-F-I-A-A-F-V-G-A-A-A-T-L-V-S-L-L-T-F-M-I-A-A-T-Y-N-F-A-V-L-
```

```
+  270  +      +
K-L-M-G-R-G-T-K-F
```

**Fig. 2.5.3:** (A) Strategy of the analysis of Proteolipid protein primary structure. (B) Dissection of PLP-sequence in hydrophobic and hydrophilic domains.

*Wilhelm Stoffel*

MBP contributes 40% of the total protein content of the CNS myelin membrane proteins whereas the main protein component (50%) is the PLP. PLP is insoluble in aqueous solution but soluble in organic solvents.

During 1980 and 1982, we unraveled the primary structure of this 276 a. a. residue containing integral membrane protein (Fig. 2.5.3A) with a combination of conventional and newly developed separation techniques. The strategy of the analysis is indicated underneath the amino acid sequence in the following figures. The polypeptide chain is extremely structured into hydrophobic and hydrophilic domains (Fig. 2.5.3B), which suggested to us the model of membrane integration shown in the Fig. 2.5.4.

Proteolysis of the hydrophilic domains exposed on the extracytosolic surface with subsequent sequence analysis of the fragments (arrows in Fig. 2.5.4) confirmed our proposal. In addition antibodies raised against syn-

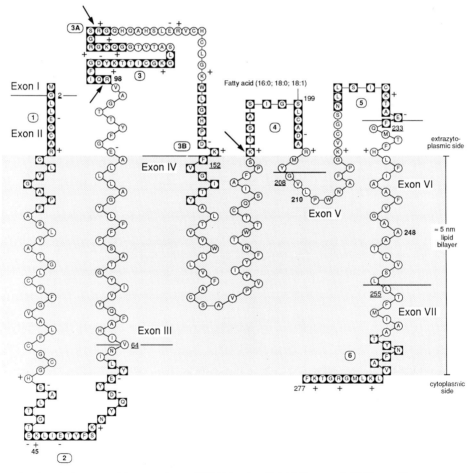

**Fig. 2.5.4:** Proposed integration of PLP in the lipid bilayer of CNS-myelin.

thetic peptides resembling the information sequences at the two sides of the bilayer (indicated as black squares in the sequences in Fig. 2.5.4) confirmed our model.

Two of the five disulfide bonds were localized on the external surface. They suggested a clustered arrangement of the *trans-* and *cis-*membranal helices as schematically drawn in Fig. 2.5.5.

In addition computer-assisted prediction of secondary structures lend strong support to an amphipathic helical arrangement of the hydrophilic loops.

Our studies suggest that the hydrophobic interactions between amphipathic helices of apposing surfaces of spirally wrapping myelin membranes leads to the compaction between membrane layers. This compaction is further supported by the intercalation of fatty acyl residues esterified to serine residues of hydrophobic extracytosolic domains of PLP. Figure 2.5.6 summarizes the three different forces leading to the compact multilayer structure of myelin.

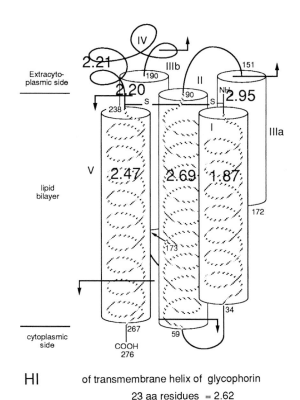

**Fig. 2.5.5:** Clustered arrangement of PLP trans- and cis-membranal helices in the lipid bilayer.

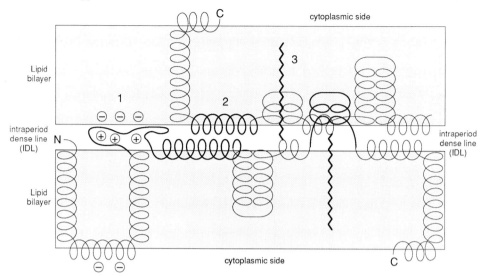

**Fig. 2.5.6:** Forces stabilizing the compact multilayer structure of myelin.

Our results regarding the primary and secondary structure and the topology of MBP and PLP are not only of great impact to the molecular architecture of this important membrane system of CNS but were also the initiation point of our molecular biological studies on this membrane.

### 2.5.3.4.2 We Conducted Molecular Biological Studies of the Myelin Membrane of CNS

We unraveled the organization and nucleotide sequences of the PLP and MBP genes. Starting with a size-fractionated cDNA library constructed from the poly(A)$^+$ RNA of myelinating rat brain (18 days) and using oligonucleotides corresponding to suitable sequences of primary structure we isolated a full length MBP and PLP cDNA clone. The nucleotide sequence-derived amino acid sequence confirmed our protein chemical data. The two cDNAs were used as probes for the isolation of the human MBP and PLP genes. Restriction mapping and nucleotide sequence analysis resulted in the gene organization shown in Figs. 2.5.7 and 2.5.8.

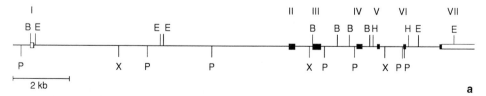

**Fig. 2.5.7:** Organisation of the human Proteolipid protein (PLP) gene.

[21] W. Stoffel, C. Schreiber, H. Scheefers: Lipids with photosensitive groups as chemical probes for the structural analysis of biological membranes. Hoppe-Seyler's Zeitschr. Physiol. Chemie 359 (1978) 923–931.

[22] W. Stoffel, P. Metz: Covalent cross-linking of photosensitive phospholipids to human serum high density apolipoproteins (apo HDL). Hoppe-Seyler's Zeitschr. Physiol. Chemie 360 (1979) 197–206.

[23] W. Stoffel, K. Preißner: Conformational analysis of serum apolipoprotein A-II in lipoprotein complexes with bifunctional cross-linking reagents. Hoppe-Seyler's Zeitschr. Physiol. Chemie 360 (1979) 691–707.

[24] W. Stoffel, E. Krüger, R. Deutzmann: Cell-free translation of human liver apolipoprotein A-I and A-II mRNA. Processing of primary translation products. Hoppe-Seyler's Zeitschr. Physiol. Chemie 364 (1983) 227–237.

[25] W. Stoffel: Synthesis, transport, and processing of apolipoproteins of high density lipoproteins. J. Lipid Res. 25 (1984) 1586–1592.

[26] W. Stoffel, H. Hillen, H. Giersiefen: Structure and molecular arrangement of proteolipid protein of central nervous system myelin. Proc. Natl. Acad. Sci. USA 81 (1984) 5012–5016.

[27] W. Stoffel, H. Giersiefen, H. Hillen, W. Schröder, B. Tunggal: Amino-acid sequence of human and bovine brain myelin proteolipid protein (lipophilin) is completely conserved. Biol. Chem. Hoppe-Seyler 366 (1985) 627–635.

[28] W. Stoffel, E. Binczek: Structural requirements of human preproapolipoprotein A-I for translocation and processing studied by site-directed mutagenesis in vitro. Biol. Chem. Hoppe-Seyler 369 (1988) 1055–1063.

[29] W. Stoffel, T. Subkowski, S. Jander: Topology of proteolipid protein in the myelin membrane of central nervous system. Biol. Chem. Hoppe-Seyler 379 (1989) 165–176.

[30] R. Streicher, W. Stoffel: The organization of the human myelin basic protein. Biol. Chem. Hoppe-Seyler 370 (1989) 503–510.

[31] H. Utsumi, B. Tunggal, W. Stoffel: Carbon-13 nuclear magnetic resonance studies on the interaction of glycophorin with lecithin in reconstituted vesicles. Biochemistry 19 (1980) 2385–2390.

[32] R. Wagener, R. Pfitzner, W. Stoffel: Studies on the organization of the human apolipoprotein B-100 gene. Biol. Chem. Hoppe-Seyler 368 (1987) 419–425.

# Doctoral Students

1974–1976:

Dr. Eckart Bauer: "Stoffwechsel von Dihydrosphingosin in Tetrahymena pyriformis." Thomae AG, Biberach a. d. Riß, FRG.

Dr. Gabriele Michaelis: "Synthese fluoreszenzmarkierter Fettsäuren und Phospholipide. Ihre Verwendung zur Aufklärung von Lipid-Lipid- und Lipid-Protein-Wechselwirkungen." Institut für Klinische Chemie, Universität Köln, FRG.

Dr. Ottfried Zierenberg: "$^{13}$C-Spin-Gitter-Relaxationszeitmessungen an Lipiden in Modell-Membranen, Lipoproteinen (High-Density-Lipoprotein) und Biomembranen (Sarcoplasmisches Retikulum)." Nattermann & Cie. GmbH, Köln, FRG.

Dr. Hans Frings: "Einfache Methoden für die Synthese von 3-sn-Phosphatidyläthanolaminen." Private Laboratory, Köln, FRG.

*Wilhelm Stoffel*

1977–1979:

Dr. Frank Bolkenius: "Untersuchungen über die Wechselwirkung des basischen (enzephalitogenen) Myelinproteins mit Sphingo- und Phospholipiden." Centre de Recherche, Strasbourg, France.

Dr. Klaus Preißner: "Topographie des Apolipoproteins AII in Lipoproteinkomplexen." Max-Planck-Institut für Klinische Forschung, Gießen, FRG.

Dr. Klaus Salm: "Synthetische $^{13}$C-angereicherte und photoaktivierbare Lipide als Sonden für Lipid-Protein-Wechselwirkungen." La Bas Pharm. Präparate GmbH, München, FRG.

Dr. Stefan Nadidei: "Einfluß der chemischen Struktur auf die Wechselwirkung zwischen komplexen Lipidmolekülen und in Lipid-Protein-Komplexen." Exxon AG, Köln, FRG.

1980–1982:

Dr. Peter Metz: "Über die räumliche Anordnung von Lipiden und Proteinen im menschlichen Serum High-Density Lipoprotein – eine Photoaffinitätsmarkierungsstudie." Boehringer AG, Mannheim, FRG.

Dr. Shailaja Javeri: "Biochemische Untersuchungen zur Asymmetrie der Membran Lipid-haltiger Viren." Institut für Immunologie, Universität Köln, FRG.

Dr. Uwe Körkemeier: "Biochemische Untersuchungen zur Asymmetrie biologischer Membranen und von Lipoproteinen." Schwartzhaupt GmbH, Köln, FRG.

1983–1985:

Dr. Heinz Hillen: "Sequenzanalyse des Gehirnmyelin-Proteolipid-Proteins – Neue Methoden zur Trennung hydrophober Peptide." BASF AG, Ludwigshafen, FRG.

Dr. Werner Schröder: Beiträge zur Primärstruktur des Folch-Proteins durch die Analyse enzymatischer Fragmente." Bayer AG, Wuppertal, FRG.

Dr. Joachim Bender: "Topographie des Apolipoproteins AI in Lipoproteinkomplexen." Bayer AG, Wuppertal, FRG.

Dr. Volker Zeugner: "Untersuchung der Proteinoberflächenstruktur des High-Density Lipoproteins mit bifunktionellen Quervernetzungsreagenzien." Behringwerke AG, Marburg, FRG.

Dr. Susanne Euler-Bertram: "$\Delta^3$-cis-$\Delta^2$-trans-Enoyl-CoA Isomerase." Institut für Physiologische Chemie, Universität Köln, FRG.

1986-1987:

Dr. Tibor Mannsfeld: "Photoaffinitätsmarkierungsstudien am menschlichen Serum High-Density Lipoprotein." Institut für Humangenetik, Universität Erlangen, FRG.

Dr. Alexander Jung: "Molekularbiologische Studien am PLP-Gen." Nattermann & Cie. GmbH, Köln, FRG.

Dr. Martin Schaich: "Myelinmembranproteine des Zentralnervensystems: Klonierung des Proteolipidproteins und des Basischen Myelinproteins aus Rattenhirn." Exxon AG, Köln, FRG.

Dr. Ralph Budzinski: "Molekularbiologische Studien der Myelinproteine." Thomae AG, Biberach a. d. Riß, FRG.

Dr. Helmut Giersiefen: "Sequenzanalyse des menschlichen Gehirnmyelin-Proteolipid-Proteins." Hoffmann La Roche, Basel, Switzerland.

Dr. Hans Josef Diehl: "Das Proteolipid-Protein-Gen des Menschen." Nattermann & Cie. GmbH, Köln, FRG.

Dr. Antje Haase: "Studien zur Expression und Regulation der Gene der Apolipoproteine AI und AII." Institut für Biochemie, Medizinische Fakultät, Universität Köln, FRG.

1988–1989:

Dr. Cordula Holtfreter: "Apolipoprotein CII. Untersuchungen zur cDNA-Expression und Funktion." Kinderklinik, Universität Düsseldorf, FRG.

Dr. Reinhard Pfitzner: "Das Apolipoprotein B 100. Molekularbiologische Beiträge zur Struktur und Funktion." Byc Santec GmbH, Dietzenbach, FRG.

Dr. Raimund Wagener: "Molekularbiologische Studien am Apolipoprotein B des Menschen." Institut für Diagnostikforschung GmbH, Berlin, FRG.

Dr. Renate Sekul: "Untersuchungen zur Prozessierung des Proapolipoproteins AI." Holphar AG, Sulzbach, FRG.

Dr. Thomas Subkowski: "Topochemische und immunologische Untersuchungen zum Proteolipidprotein." BASF AG, Ludwigshafen, FRG.

# Postdoctoral Scientist

1979–1983:

Dr. Rainer Deutzmann, Institut für Biochemie, Regensburg, FRG.

# 2.6 Resolution of Recombination Intermediates by Endonuclease VII *in vitro*

Börries Kemper*

## 2.6.1 Summary

The analysis of DNA-packaging defective T4 mutants in gene 49 led to the discovery, isolation and finally the characterization of Endonuclease VII (Endo VII). The specificity of the enzyme for DNA secondary structures and its potency to accurately resolve Holliday-structures *in vitro* made it a prototype of a new class of nucleases, which we named **X-solvases** because of their ability to cleave cruciform DNAs (X-structures). The search for more enzymes of this type led to a few already well-known recombinases and several new isolates from different prokaryotic sources like phages T7, Pl, lambda, and *Escherichia coli*, or from eukaryotic sources like *Saccharomyces cerevisiae*, HeLa cells, and human placenta (see Table 2.6.1).

## 2.6.2 Preface

The initial aim of my work was the isolation and characterization of enzymes (nucleases) involved in genetic recombination of bacteriophages and their hosts. Two ways seemed practicable at that time: (1) Random screening for any nuclease followed by a search for related genes and characterization of mutants or (2) starting from mutants in known genes and searching for the appropriate enzymes in the wildtype organism. In 1972, when the methods of genetic engineering were not yet available, the second approach seemed more practicable and was, therefore, chosen. Elaborate work from the laboratory of F. Frankel on mutations in gene 49 of bacteriophage T4 [1] had led to a description of a very interesting phenotype (to be described below),

---

* Institut für Genetik der Universität zu Köln

**Table 2.6.1:** X-Solvases (as per January 1991).

| Enzyme | Source | Gene |
|---|---|---|
| Endo VII | Phage T4 | 49 |
| Endo I | Phage T7 | 3 |
| cre | Phage Pl | cre |
| Exo V | *E. coli* | recBCD |
| No name | *E. coli* | ruvC |
| Int | Phage lambda | int |
| EndoX 1 | *S. cerevisiae* | ? |
| EndoX 2 | *S. cerevisiae* | ? |
| EndoX 3 | *S. cerevisiae* | ? |
| EndoX 4 | *S. cerevisiae* | cce1 |
| FLP protein | 2μm plasmid | flp |
| No name | HeLa cells | ? |
| No name | Human placenta | ? |
| No name | calf thymus | ? |

which directed my attention to a yet unexplored aspect of genetic recombination: the resolution of recombination intermediates, in particular branched DNA structures. During the following years, this problem became the major issue of the work in my laboratory.

## 2.6.3   Introduction

The model of recombination, originally proposed by R. Holliday [2], requires exchanges between strands of like polarity after synapsis of two double-stranded DNA molecules is established (Fig. 2.6.1). The resulting crossover leads to the Holliday-structure as the central intermediate. This structure has the ability to 'branch migrate' into either direction, away from the original site of exchange, thereby extending the region of heteroduplex DNA. During this time, isomerization can occur which now allows those strands, which were not broken and exchanged in the first place, to cross over. It was assumed that resolution of a Holliday-structure occurs at crossing strands, affecting the two pairs of homologs with an equal probability. Successful resolution would then lead to genetically distinguishable "patch"- and "splice"-recombinants. The resolution of Holliday-structures is, therefore, a key event during genetic recombination and was assumed to be enzymatically driven [2].

## Homologous Genetic Recombination (model)

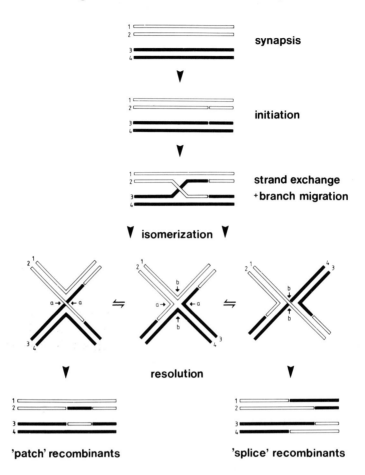

**Fig. 2.6.1:** Model of genetic recombination. The figure shows different stages of homologous genetic recombination as proposed by Robin Holliday [2].

For a frequently recombining organism like phage T4, reduction or loss of the ability to properly resolve Holliday-structures was expected to severely affect several stages of the development. Beside a general impairment of exchanges between genetic markers, replication should also be affected, since it is dependent on recombination in this organism, and the packaging of newly synthesized DNA should also be disturbed because it depends on unbranched and concatenated molecules. Mutations in genes 49, 16, and 17 were originally classified as morphogenic mutations, exhibiting DNA-packaging defects [3]. However, mutations in gene 49 differed from mutations in the other two genes by showing partially filled heads in the electron microscope [4], suggesting a defect in the processing rather than the initiation of packaging events [5]. The defect could principally affect the heads or the DNA. To pursue F. Frankel's results, we decided to analyze the unpackagable DNA from gene 49⁻-mutants first.

## 2.6.4   Isolation and Characterization of VFS-DNA

The 49⁻-DNA accumulates in very fast-sedimenting complexes (VFS-DNA). Our early attempts to isolate this DNA involved the M-band technique adopted from M. Schaechter's laboratory [6], which was based on the observation that crystals of magnesium sarkosynate bind tightly to newly replicated DNA (according to Schaechter, presumably via membrane fragments). To our surprise the method worked efficiently with 49⁻-DNA, though this DNA was completely free from any membrane components, and we were able to isolate large quantities of VFS-DNA for use in biochemical studies [7, 8]. Inspection of the DNA in the electron microscope revealed highly compact complexes with domain-like loops, some of them in a super-coiled state (Fig. 2.6.2 a). Determination of the exact number of complexes in solution and the total amount of DNA suggested that each complex contained approximately 280 to 320 T4-unit chromosomes, which is equivalent to a whole burst. After cross-linking the DNA to protein by formaldehyd followed by purification of the products over a sucrose gradient, we provided evidence that in gene 49⁻-mutants newly synthesized phage heads become attached to free DNA-ends, which protrude from the complexes at all stages of incomplete filling [8]. Many DNA-branches were found immediately in front of partially filled heads. This finding supports the hypothesis that gene 49 was responsible for converting unpackagable branched DNA into packagable unbranched DNA. As a side product of these investigations, it became apparent that packaging *in vivo* involves many DNA-termini at the same time (Fig. 2.6.2b). Details of the complexity of VFS-DNA became visible when mutations in gene 49 were combined with mutations in other genes (T4-strain GT7–49⁻), which replace hydroxymethylcytosine by cytosine making T4-DNA susceptible to restriction enzymes. Since the replacement had no influence on the described characteristics of VFS-DNA, we digested the complexes into smaller units with ClaI and analyzed sucrose gradient-purified fragments with the electron microscope. An unusually high number of Y- and X-shaped DNA structures was detected in the DNA (Figs. 2.6.2c and 2.6.2d).

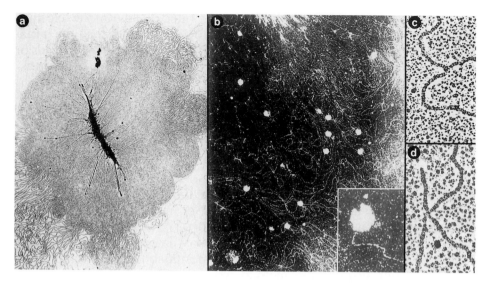

**Fig. 2.6.2:** Very fast-sedimenting DNS (VFS-DNA) from 49⁻-infections. The sequence of pictures shows different stages of the analysis of VFS-DNA isolated from gene 49 defective phage infected-cells. (a) Purified VFS-DNA visualized in the electron microscope. (b) VFS-DNA after crosslinking with formaldehyd, showing partially filled heads attached to free ends. The inset shows a magnification of a head structure during filling with DNA. (c, d) Two branched DNA structures isolated from cytosine containing VFS-DNA after digestion with ClaI. Examples of Y- and X-structures are shown.

## 2.6.5 Isolation and Characterization of Endonuclease VII

If gene 49 would code for a nuclease responsible for 'unbranching' newly synthesized DNA, it should be possible to isolate the activity by using VFS-DNA as a substrate *in vitro*. We measured the degradation of purified VFS-DNA after addition of crude extracts from phage T4-infected cells *in vitro* by using an assay based on the above described M-band technique. The assay allowed a quick separation of intact undigested VFS-DNA from slow sedimenting fragments on mini-sucrose-gradients (1 ml) by centrifugation in a Sorvall centrifuge. Comparison between extracts from T4 wildtype-infected cells with extracts from gene 49⁻-phage-infected cells opened the way to the isolation of a VFS-DNA degrading activity [9, 10]. In continuation of an already existing classification of T4 induced nucleases [11] we named the activity Endonuclease VII (Endo VII) [9].

Gene 49 was recently cloned and sequenced [12, 13], and it shows an open reading frame (ORF) for an 18 kD protein. The gene was overexpressed in *E. coli* [14] and induced clones were used for purification of large quantities of cloned Endo VII (Endo VIIcl). The pure (» 99 %) protein has the predicted molecular weight of 18 kD on denaturing PAGE, while under native conditions, a MW of 38 kD was determined [15]. We believe that the enzyme works as a homo-dimer on branched DNAs, and a working model of its presumed action is presented at the end of this report.

## 2.6.6    Endonuclease VII Resolves Holliday-Structures

Proof for the ability of Endo VII to resolve Holliday-structures *in vitro* was obtained from a collaboration with colleagues from the NIH in Bethesda (USA), who had isolated recombinant DNA from *in vivo* reactions, so-called "figure-8-structures" and "α-structures". The Holliday-structure resolving activity of Endo VII was directly demonstrated by analyzing the branched DNA in the electron microscope before and after treatment with the enzyme [16].

Since it is difficult to isolate Holliday-structures from *in vivo* recombination events in large quantities and since these structures, due to unrestricted branch migration, are never in a fixed position on the chromosome, they were not useful for a more detailed analysis of Endo VII reactions. It was an important observation when Endo VII was found to be capable of resolving cruciform structures at palindromes in supercoiled DNA [17, 18]. Endo VII cleaved these structures at the base of the hairpins by placing pairs of cuts across the junction. Cuts, releasing the short arms (hairpins) from the structure, were not found. The products from reactions with Endo VII had the short arms connected to either side of linear molecule in a telomere-like conformation exhibiting ligatable nicks between the fold-back structure and the remainder of the molecule. We have recently shown, that Endo VII is also active on fully synthetic cruciforms (Fig. 2.6.3), which were made *in vitro* by annealing four complimentary synthetic oligonucleotides [19]. Since one can formally consider cruciforms as a sort of Holliday-structures (viewing them as X-structures), we will be able to study the biochemistry of resolution of recombinant intermediates *in vitro* with these substrates. We are aware of the fact that in a 'true' Holliday-structure, two pairs of homologous arms are linked by the junction, while in a supercoiled cruciform structure, there is only one pair of homologous arms joined with one pair of non-homologous arms. A synthetic cruciform is bare of any homology. Since Endo VII does not discriminate between these structures, we have suggested to give this class of enzymes the name X-solvase(s), where the "X" stands as a symbol for crossing strands [19].

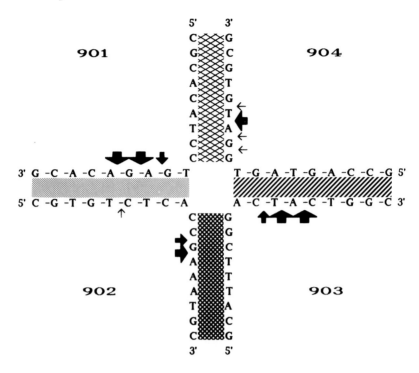

**Fig. 2.6.3:** Endo VII induced cleavage patterns in cruciform DNA. Cruciform DNA was made by *in vitro* hybridization of four synthetic oligonucleotides. The arrows identify locations of cleavage sites. The size of the arrows indicates the relative usage of sites. Note that the arms of the structure have no homologies, which is indicated by different shadings.

## 2.6.7 Endonuclease VII Cleaves Secondary Structures in DNA

Since Endo VII did not distinguish between true Holliday-structures and cruciform DNA, we were curious to see whether the enzyme would recognize DNAs with other secondary structures as well. By hybridizing plus- and minus-strands from suitable derivatives of phage Ml3mp2 we created a series of circular full-sized (7.200 bp) molecules with Y-branches and heteroduplex loops. Both structures were equally well recognized and cleaved by Endo VII [20, 21]. Clusters of cleavages were found located 3' from the structure in both stands (results with heteroduplex loops are shown in Fig. 2.6.4a). The

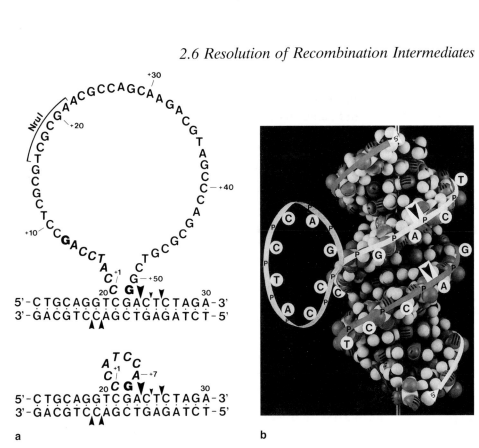

**Fig. 2.6.4:** Endo VII cleaves at heteroduplex loops. Heteroduplex loops of defined sequence and size were made *in vitro* by hybridizing the plus- and minus-strands from phage M13mp2 derivatives. (a) Sections forming the loop of 51 nt and 8 nt, respectively, are shown here. Arrows point to Endo VII-induced cleavage sites. The size of the arrows reflects relative usage of the indicated positions for cleavage by Endo VII. (b) Three dimensional model of a piece of loop-DNA, showing that pairs of cuts located 3' from the loop in the looping strand or 3' from the base of the loop in the non-looping strand are located on the same side of the molecule at close distance and reachable across the minor groove. 3'- or 5'-end labelled substrates were treated with Endo VII and the cleavage sites were mapped by appropriate techniques. The arrows point to Endo VII cleavage sites.

cleavage patterns were reproducible but varied with different structure suggesting an influence of local nucleotide sequence. By using a 'simple' substrate for Endo VII, which is a short double-stranded oligonucleotide with a single-stranded overhang, we systematically changed the sequences at and around regular cleavage positions and analyzed the influence on locations and usage of cleavage sites. A rule became apparent: Pyrimidines are preferred to purines and the actual nucleotide sequence at the structure affects the cleavability of regular sites (S. Pottmeyer, in prep.).

What is the smallest DNA secondary structure cleavable by Endo VII? We created a set of substrates from synthetic oligonucleotides by hybridizing two, three or four molecules with limited complementarity and treated the resulting branched substrates with Endo VII (S. Pottmeyer, in prep.). The enzyme always cleaves 3' from the site where one strand either swaps its complementary partner or continues in a single-stranded tail or loop.

We want to emphasize that Endo VII does not cleave single-stranded DNA, neither linear molecules nor looping strands. If, however, cleavage of "single-stranded DNA" is observed, the presence of secondary structures can be predicted. The enzyme might, therefore, be useful in analyzing DNA secondary structures (S. Pottmeyer, in prep.).

## 2.6.8   Endo VII Cleaves at Mismatches

In continuation of the investigations described in 1.6, we also investigated whether the enzyme acts on mismatches. Double-stranded DNAs with different mismatches were made from oligonucleotides by the above described procedure, and we found that the enzyme acted in response to base mismatches, cleaving the DNA 3' from the obstruction in both strands. This causes linearization of the oligonucleotide by pairs of staggered nicks (P. Solaro, in prep.). The resulting double-strand break, however, does not seem biologically meaningful, and we believe that *in vitro* the reaction of the purified enzyme goes too far. Attempts are in progress to look for factors from phage infected cells, which might regulate the described cleavage events, for example by reducing counternicking activity.

## 2.6.9   Purification of Endonuclease X3 from Yeast *Saccharomyces cerevisiae*

Despite the persuasiveness of the *in vitro* studies, we are still lacking convincing evidence that Endo VII is indeed required for the resolution of Holliday-junctions *in vivo*. Measurements of recombination frequencies between rII-markers in the presence and absence of a functional gene 49 did not reveal a factor greater than two, suggesting that Endo VII contributes to recombination but might not be the major component in T4 recombination pathways. We, therefore, wanted to know whether Endo VII is just a specialized phage enzyme or whether other organisms have similar activities.

With an assay procedure based on the cleavability of cruciform DNA in supercoiled DNA, we screened crude extracts from yeast *Saccharomyces cerevisiae* and found an activity which was named Endonuclease X3 [19], since two other groups had also described X-solvase activities from yeast after using the same assay [22, 23]. Endonuclease X3 was tested with a series of DNA substrates, and a surprising similarity in the cleavage patterns between Endo VII and this enzyme was found. In addition, Endo X3 showed the same molecular weight of 18 kD on SDS PAGE and approximately 40 kD under native conditions as Endo VII did. Furthermore, Endo X3 is sensitive towards an anti-Endo VII antiserum, which does not inhibit, for example, Endonuclease I from phage T7 or a collection of various restriction enzymes [15]. Comparison of the properties of Endo X1 [22], Endo X2 [23], and Endo X3, however, revealed marked differences among the three enzymes, suggesting that Endo X3 is not related to any one of the other two enzymes. A final answer, however, has to await isolation of the genes. This work is in progress in our's and the other laboratories.

## 2.6.10  Collaborations

The potential of Endo VII for studying DNA secondary structure was recognized and stimulated the work of several groups working with synthetic cruciforms or Holliday-structure analogs. In collaboration with two other groups, it was independently found that synthetic cruciforms show a two-fold symmetry, rather than a four-fold symmetry. Both studies suggest that the arms are aligned pairwise and appear preferentially locked in either of the two isomeric forms. The decision, which of the two isomers will be adopted, is influenced by the nucleotides located immediately at the junction [24, 25]. If one applies these results to the situation *in vivo*, one can envisage how, during branch migration, the crossing strands are continuously changing, while the structure flips over. This should slow down the process of branch migration. Recent investigations have indeed shown that branch migration is a comparably slow process, much slower than one would have expected from theoretical calculations.

In another collaboration, S. West tried to simulate a series of events of homologous recombination in an *in vitro* system, involving pairing and strand exchanges driven by the recA protein and resolution of the resulting Holliday-junctions by Endo VII. The process does indeed go to completion and individual steps can now be studied in greater detail [26].

## 2.6.11  The Model

Looking back to the initially described situation where the generation of recombinant products from the Holliday-junction was described in accordance with the Holliday-model (Fig. 2.6.1), it seems plausible that an X-solvase activity would bind to two of the homologous arms and cleaves the crossing strands on equal distance from the crossover-site. A homodimer of Endo VII could make these cleavages (Fig. 2.6.5), and we are proposing a 3-step model as a working hypothesis:

**Step 1.**  The enzyme searches the DNA by sliding in a 3' to 5' direction on one strand of the molecule. The length of sliding paths is shorter than 9 nt because no change in rate or yield of cleavage at a hetero-duplex loop was observed when the arm length was varied between 7200 nt (a whole genome of M13) and 9 nt. Cruciforms with 8 nt

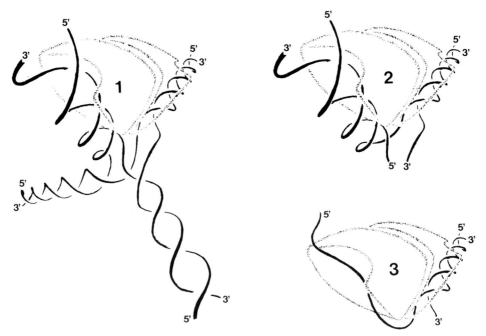

**Fig. 2.6.5:** Model of Endo VII action. The sketches show three different DNA-substrates (X-DNA, nicked double-stranded DNA, and double-stranded DNA with a single-stranded overhang). Endo VII cleaves each of these substrates efficiently, and two subunits of the enzyme are drawn in hypothetical locations on the substrates. Close contact of each subunit to different portions of the DNA-molecules induces bends in the DNA.

long arms, however, were poor substrates in comparison to cruci-forms with 9 nt (Jensch and Seeman, unpublished observations). The weak binding of Endo VII to its substrates would be in agreement with short sliding tracks.

**Step 2.** The enzyme gets stuck after hitting an obstruction in the DNA and starts cleaving 3' from this site at a minimum distance of 2 nt. The efficiency of cleavage and the distribution of cleavage sites depends on the local sequence (pyrimidines being preferred), and the obstruction can be anything from a single unpaired basepair (mismatches, single-nucleotide additions) to branches in one or both strands or even single-strand overhangs. Recently it was shown that bent or kinked DNA was also a substrate for Endo VII (D. Lilley, personal communication).

**Step 3.** The enzyme is a homo-dimer with two binding-sites and two active centers. When one subunit is docked to its position for cleavage, the second subunit gets into contact with the DNA. If the obstruction is a branch, the enzyme will reach for the other arm and due to the geometry of the structure and the order of the enzymes's subunits, the correct arm will be cleaved in the appropriate position to give precise and ligatable nicks, provided both strands have the same sequence due to homology. If there is no homology, each arm will be cleaved with characteristically different cleavage patterns, which are directed by local sequences. In this situation it might happen that two combined nicks do not produce ligatable nicks, since a short gap or nucleotide-overlap is generated at the termini.

Why does the enzyme cleave at nicks or mismatches, if there is no branch in the DNA? Following the above described idea, we can envisage how the enzyme gets stuck and trapped by bends or kinks in the DNA, which are formed at structural deviations. Additional cleavage of the opposite strand of the same molecule could occasionally occur as a mistake. We have observed that opening a nick into a gap provides a better substrate for Endo VII, probably by reducing the stacking forces between the two nucleotides flanking the nick and giving the molecule a greater flexibility to form sharper bends. This hypothesis can be tested. A similar situation is found at junctions between double-stranded and single-stranded DNAs, where the increasing length of the singlestranded overhang gives the enzyme a better halt by direct contact, as illustrated in Fig. 2.6.5. We want to point out that Endo VII cleavages will be located on the same side of a DNA molecule if one looks at a three dimensional model, e. g. from a heteroduplex loop (Fig. 2.6.4b). This might be a clue to an understanding of the reaction mechanism of Endo VII with its substrates [27].

93

*Börries Kemper*

## 2.6.12 References

(x = own publications, o = publications from collaborations)

[1] F. R. Frankel: Evidence for long strands in the replicating pool after T4 infection. Proc. Natl. Acad Sci. USA 59 (1968) 131–138.
[2] R. Holliday: A mechanism for gene conversion in fungi. Genet. Res. 5 (1864) 282–304.
[3] R. B. Luftig, C. Ganz: Phage head morphogenesis. IV Comparison of gene 16-, 17- and 49-defective head structures. J. Virol 9 (1972) 377–389.
[4] R. B. Luftig, W. B. Wood, R. Okinaka: Phage head morphogenesis. On the nature of gene 49-defective heads and their role as intermediates. J. Mol. Biol. 57 (1971) 555–573.
[5] U. K. Laemmli, N. Teaff, J. D'Ambrosia: Maturation of the head of phage T4. III DNA packaging into preformed heads. J. Mol. Biol. 88 (1974), 749–765.
[6] C. F. Earhart, G. Y. Tremblay, M. J. Daniels, M. Schaechter: DNA replication studied by a new method for the isolation of cell membrane-DNA complexes. Cold Spring Harb. Symp. Quant. Biol. 33 (1968) 707–710.
x [7] B. Kemper, E. Janz: Function of gene 49 of bacteriophage T4. I. Isolation and biochemical characterization of very fast-sedimenting DNA. J. Virol. 18 (1976) 992–999.
x [8] B. Kemper, D. T. Brown: Function of gene 49 of bacteriophage T4. II. Analysis of intracellular development and the structure of very fast-sedimenting DNA. J. Virol. 18 (1976) 1000–1015.
x [9] B. Kemper, M. Garabett: Studies on T4 head maturation. I. Purification and characterization of gene-49-controlled endonuclease. Eur. J. Biochem. 115 (1981) 123–131.
x [10] B. Kemper, U. Courage, M. Garabett: Studies on T4 head maturation. 2. Substrate specificity of gene-49-controlled endonuclease. Eur. J. Biochem. 115 (1981) 133–141.
x [11] B. Kemper, J. Hurwitz: Studies on T4-induced nucleases. Isolation and characterization of a manganese-activated T4-induced endonuclease. J. Biol. Chem. 248 (1973) 92–99.
[12] K. A. Barth, D. Powell, M. Trupin, G. Mosig: Regulation of two nested proteins from gene 49 (recombination endonuclease VII) and of a λ rexA-like protein of bacteriophage T4. Genetics 120 (1988) 329–343.
[13] J. Tomaschewski, W. Rüger: Nucleotide sequence and primary structures of gene products coded for by the T4 genome between map positions 48.266 kb and 39.166 kb. Nucl. Acid. Res. 15 (1987) 3632–3633.
[14] J. Tomaschewski: Endonuklease VII des Bakteriophagen T4: Sequenzierung und Überexpression. PhD Thesis, Universität Bochum, FRG 1988.
x [15] H. Kosak, B. Kemper: Large-scale preparation of T4 endonuclease VII from over-expressing bacteria. Eur. J. Biochem. 194 (1991) 779–784.
o [16] K. Mizuuchi, B. Kemper, J. Hays, R. Weisberg: T4 endonuclease VII cleaves Holliday structures. Cell 29 (1982) 357–365.
o [17] B. Kemper, F. Jensch, M. v. Depka-Prondzynski, H.-J. Fritz, U. Borgmeyer, K. Mizuuchi: Resulotion of Holliday structures by endonuclease VII as observed in interactions with cruciform DNA. Cold Spring Harbor Symp. Quant. Biol. 49 (1984) 815–825.

o [18] D. M. J. Lilley, B. Kemper: Cruciform resolvase-interactions in supercoiled DNA. Cell 36 (1984) 413–422.

x [19] F. Jensch, B. Kemper: Cruciform cutting endonuclease from *Saccharomyces cerevisiae* and phage T4 show conserved reactions with branched DNAs. EMBO J. 8 (1989) 4325–4334.

x [20] F. Jensch, B. Kemper: Endonuclease VII resolves Y-junctions in branched DNA *in vitro*. EMBO J. 5 (1986) 181–189.

x [21] S. Kleff, B. Kemper: Initiation of heteroduplex repair by T4-encoded endonuclease VII *in vitro*. EMBO J. 5 (1988) 1527–1535.

[22] S. C. West, A. Körner: Cleavage of cruciform DNA structures by an activity from *Saccharomyces cerevisiae*. Proc. Natl. Acad. Sci. USA 82 (1985) 6445–6449.

[23] L. S. Symington, R. Kolodner: Partial purification of an enzyme from *Saccharomyces cerevisiae* that cleaves Holliday junctions. Proc. Natl. Acad. Sci. USA 82 (1985) 7247–7251.

o [24] D. R. Duckett, A. I. H. Murchie, S. Diekmann, E. v. Kitzing, B. Kemper, D. J. M. Lilley: The structure of the Holliday junction, and its resolution. Cell 55 (1988) 79–89.

o [25] J. E. Mueller, B. Kemper, R. P. Cunningham, N. R. Kallenbach, N. C. Seeman: T4 endonuclease VII cleaves the crossover strands of Holliday junction analogs. Proc. Natl. Acad. Sci. USA 85 (1988) 9441–9445.

o [26] B. Müller, C. Jones, B. Kemper, S. C. West: Enzymatic formation and resolution of Holliday junctions in vitro. Cell 60 (1990) 329–336.

x [27] B. Kemper, S. Pottmeyer, P. Solaro, H. Kosak: Resolution of DNA-secondary structures by endonuclease VII (Endo VII) from phage T4. In: R. H. Sarma, M. H. Sarma (eds.): Structure and Methods Vol. 1. Adenine Press, Schenectady, N. Y. 1990, pp. 215–229.

## Diploma Theses

U. Courage (1981): Untersuchungen zur Spezifität der T4-induzierten Endonuklease VII. Reaktion des Enzyms mit partiell einzelsträngigen DNAs.

H. Hoffmann (1983): Untersuchungen zur Substratspezifität von Endonuklease VII. Konstruktion ,Cruciformer' DNA *in vitro*.

M. v. Depka-Prondzynski (1984): Feinkartierung der Endonuklease VII induzierten Schnittstellen in cruciformer DNA.

F. Jensch (1984): Untersuchungen zur Substratspezifität von Endonuklease VII: Charakterisierung der Schnittstellen in partiell einzelsträngiger oder verzweigter DNA.

S. Pottmeyer (1985): Untersuchungen zur Substratspezifität von Endonuklease VII. Abbau einzelsträngiger DNA.

S. Kleff (1987): Herstellung von Heteroduplex-DNA *in vitro* und ihre Auflösung durch Endonuklease VII.

M. Serwe (1987): Nachweis und Vergleich von Sekundärstrukturen im Bereich der Replikationsstartstellen der Phagen Ike und Ml3 mit Hilfe von Endonuklease VII.

H. Neuhaus (1989): Erkennung von DNA-Strukturen durch Endonuklease VII. Beispiele: Paranemische DNA und Z-DNA.

P. Solaro (1989): Reaktionen von Endonuklease VII mit synthetischen Heteroduplex DNAs.

*Börries Kemper*

## Doctoral Theses

F. Jensch (1989): Nachweis, Reinigung und Charakterisierung einer DNA-Struktur-spezifischen Endonuklease (Endonuklease X3) aus *Saccharomyces cerevisiae.*

S. Pottmeyer (1989): Sequenz- und Strukturspezifität der Endonuklease VII des Bakteriophagen T4. *In vitro* Untersuchungen an synthetischen DNA-Strukturen.

96

# 3    Plant Molecular Biology

# 3.1 Transposable Elements in Bacteria and in Plants[1]

Heinz Saedler* and Peter Starlinger**

## 3.1.1 Summary

The work on transposable elements in Köln began before the Sonderfor-schungsbereich 74 was founded, and it is continuing beyond its duration. Some of it might have become possible by funding through other sources, e. g. the „Normalverfahren" of the Deutsche Forschungsgemeinschaft. We have no doubt, however, that the existence of this Sonderforschungsbereich not only made the task of securing long-term funding for this research much easier, but we also benefited very much by the continued discussions with our colleagues who were attracted to Köln by the existence of the Sonder-forschungsbereich or else did not follow interesting offers to other places because of the existence of the research environment in Köln.

On our discussion of transposon research in Köln, however, we will not restrict ourselves closely to the time during which we obtained our funds through the Sonderforschungsbereich.

## 3.1.2 The Discovery of IS-Elements
(Heinz Saedler and Peter Starlinger)

IS-elements (insertion sequences) in *E. coli* were discovered simultaneously by J. Shapiro, who worked first in London and later in Paris and by our laboratory in Köln, where we were joined in our work by Elke Jordan, who is now at the National Institutes of Health, where she has just started to manage the

---

[1] Dedicated to the memory of Max Delbrück and Josef Straub and to Carsten Bresch who founded the Institut für Genetik

\* Max-Planck-Institut für Züchtungsforschung, Köln-Vogelsang
\*\* Institut für Genetik der Universität zu Köln

office of the human genome sequencing project. In both laboratories, the discovery of transposable elements was unintended.

The aim of both laboratories was the study of polar effects in bacterial operons, which was an interesting research topic in the early 1960's.

In order to isolate such mutations, we, as well as Shapiro, employed the selection against active galactokinase in mutants unable to metabolize galactose-1-phosphate, and therefore, were sensitive to the accumulation of this product. This accumulation could be either avoided by a mutation of the kinase gene itself or by a polar mutation earlier in the operon.

Such mutants were easily obtained, and we began to study them by the usual methods of bacterial genetics. This led to a surprise. The mutants had a high reversion rate to wildtype. Thus, they could not be deletions. On the other hand, these mutagens could not be increased by a number of mutagens known to cause either base substitutions or frameshift mutations. This made it unlikely that our mutants were caused by these alterations of DNA structure.

Thus, we had to look for other types of mutations, e. g. chromosomal aberrations. These had not been well described in bacteria by this time, and also it was not clear which methods would be suitable to detect them in those days, when cloning and sequencing was not yet available.

As deletions seemed to be excluded, duplications, inversions, or translocations well-known from eukaryotic cytogenetics remained as a possibility. A test for an inversion would be difficult because the genetic map was not dense enough to easily allow the detailed mapping experiments suitable to show an altered order of genetic markers. Therefore, we decided to first test two other possibilities, namely the presence of additional DNA that could be caused by either a duplication or a translocation, and subsequently, the study of the additional DNA in order to show whether this additional DNA, if present, was a duplication of sequences belonging to the galactose operon, or whether new DNA sequences were inserted there.

How could the presence of additional DNA be detected? We would have to investigate a DNA molecule of known length and ask whether this length increased in the known mutants. This posed two different problems. How could we obtain a segment of DNA of known length from the large *E. coli* chromosome. Transducing lambda phages offered themselves. Each of the phages is characterized by the loss of a defined segment of lambda DNA and by the uptake of a defined segment of bacterial DNA. These defective transducing phages can be propagated in the presence of wildtype phage, and thus, large amounts of DNA molecules of identical length can be isolated.

Genetic recombination between the galactose operon carried by the transducing lambda phage and the mutated galactose operon carried by the chromosome of the mutant bacteria would transfer the mutation to the known phage, which would thus be isogenic with its progenitor but for the presence of the mutation. The two phages could then be compared.

100

The length of the DNA could have been measured in the electron microscope or by the sedimentation in a sucrose gradient, but given the length of the molecule and not knowing whether an additional DNA segment, if present, would be large, we did not consider these methods to be sensitive enough. Heteroduplex mapping was not yet invented. We thus decided to use the very sensitive equilibrium centrifugation in caesium chloride. This was possible because of the then well-known fact that the density of a bacteriophage was determined by the average of the relative contributions of its protein of low density and its DNA of high density. When the protein was constant, an alteration in the amount of DNA should translate itself into a measurable alteration in the density of the two phages in the same tube.

In order to avoid absolute measurements, we compared the phages by labelling with different radioisotopes. The result was quite clear: The mutant carried an additional amount of DNA that we estimated to be in the order of between one half and one kilobase, meaning an increase in the DNA amount of the phage by approximately one to two percent. In order to confirm this, we selected revertants to the wildtype phenotype on the phage and subjected them to CsCl-centrifugation. These revertants banded indistinguishable from wildtype. Shapiro independently found the same result at the same time and, as the two studies confirmed each other, we could take this result for granted, though the effect was small.

As a next step, we tried to distinguish between duplications and translocations. We chose RNA-DNA hybridization. We transcribed mutant DNA *in vitro* with *E. coli* RNA polymerase (in a very naive manner, hoping that every sequence would be transcribed regardless of the presence of promoters, etc.). This RNA was exhaustively hybridized to DNA of the isogenic wildtype, and it was then asked whether a specific fraction hybridizing only to mutant DNA would remain. This was indeed the case. We took this as evidence for translocation of material that came from outside of the DNA carried by the bacteriophage.

In these experiments, it turned out that the insertion-specific DNA was the same in two different mutants. This opened the possibility that the foreign DNA found in different independent mutants was not randomly translocated there but was rather belonging to specific classes. We followed this up using the newly introduced techniques of heteroduplex mapping and were indeed able to show that the mutants isolated by us belonged to only a few different classes. Collaborations with Phillip Brachet from Paris, and with Waclaw Szybalski and Michael Malamy from the USA showed that these insertion classes showed up in independent sets of mutants obtained in bacteriophage lambda and the lactose operon, respectively. The latter mutants had been isolated by Malamy even before our mutants in the galactose operon. They had not been interpreted in our way because they responded to frameshift mutagens in a not yet understood way.

By this time it dawned on us that we were seeing something similar to the transposable genetic elements that had been described long ago by Barbara McClintock. This late realization was not due to the fact that we had not been familiar with McClintock's work. We had often discussed it in seminars because we found her experiments very interesting. We did not draw the parallel to bacteria, however, before it became clear that we were seeing preformed elements rather than random transpositions.

At this time, the laboratory of Heinz Saedler was separated from that of Peter Starlinger, and the former left Köln shortly after to become an Associate Professor at the University of Freiburg without stopping the continuous and frequent exchange of ideas that led to the publication of joint review articles.

It was during the former period of collaboration that the Sonderforschungsbereich was founded in 1970, and some of the work described up to now was already funded through this source.

## 3.1.3    Studies on IS-Elements in Köln
### (H. Saedler)

After the IS-elements had been shown to be discrete DNA entities we began to study various properties of IS1 and IS2.

Burning questions in these days were related to the origin of these elements i. e. are IS'es constituents of the *E. coli* chromosome or rather infectious agents.

*IS-elements are normal constituents of the E. coli chromosome and of some plasmids*

We attempted to solve this question by DNA hybridizations. This was not such an easy task since the Southern blotting technique was not yet developed. Therefore an IS-element was loaded on small ($\varnothing$ 5 mm) filters and hybridized to *in vivo* labelled *E. coli* DNA sheared by sonication to small (500 bp) long fragments. As controls lambda DNA not carrying insertion elements were used, such that the difference between these hybridizations indicated the amount of IS DNA sequences present in the labelled DNA material. The system was properly calibrated for quantitation and the results were striking. We observed about 8 copies of IS1 and about 5 copies of IS2 as chromosomals components of the *E. coli* strains used. Many years later other authors confirmed these experiments using Southern blotting techniques.

During the experiments in 1973 we noticed that IS1 and IS2 are not only natural components of the *E. coli* chromosome but also of some of its epi-

supported and extended mutation studies mentioned in the last paragraph.

# 3.1.6 Studies on Plant Molecular Biology in Köln
## (H. Saedler)

In 1980 when I returned to Köln to join the Max-Planck-Institut für Züchtungsforschung I also became a member of the Sonderforschungsbereich 74 again. However, our work on bacterial IS-elements had been terminated and rather experiments with plants were carried out in a variety of areas:

a) Transposable elements of *Antirrhinum majus* (Snapdragon) and *Zea mays*
b) Regulation of the anthocyanin biosynthetic pathway in maize
c) Plant transformation systems
d) Evolution of Angiosperms
e) Flower development of *A. Majus*

Since the Sonderforschungsbereich funded some work concerning transposable elements in *A. majus* I will describe this in particular even though our genetic and molecular understanding of the En/Spm element of *Z. mays* is much deeper.

*Paramutation in Antirrhinum majus*

The phenomenon of paramutation is widely spread in plants and in *A. majus* it was amenable to molecular analysis. This was made possible through support by the Sonderforschungsbereich 74.

   Paramutation in *Antirrhinum majus* might be due to interactions of different though related transposable elements. At the beginning of the eighties we had isolated the nivea locus from *A. majus* encoding chalcone synthase (chs) an enzyme involved in anthocyanin biosynthesis. An unstable mutant at that locus isolated in the thierties by Kuckuck turned out to be due to the integration of a 15 kb large transposable element which we called Tam1. This element is integrated in the promoter region of the chs gene. It reverts at a high rate, i.e. 10–20 % of the progeny is revertant. This instability is also observed somatically resulting in a highly variegated flower phenotype: red spots and stripes on an almost colorless background. If this mutation, termed niv-53, is crossed to a stable white nivea mutant 100 % of the progeny are variegated as is expected. However, if niv-53 is crossed to one particular stable white nivea mutant, niv-44, a mixture of F1 phenotypes is seen, which most importantly give no variegated F2 progeny. All F2s are white. It is this lack of segregation which qualifies the above cross as paramutagenic. Niv-53 is said to be paramutable, while niv-44 is paramutagenic.

Formally the former allele undergoes a mutation under the influence of the latter (niv-44) into the direction of the latter allele. Paramutation thus seems to be *in vivo* directed mutagenesis.

This phenomenon has best been described genetically at the R locus of maize, but also exists in *A. majus* as shown above and most importantly could be analyzed here because of the molecular availability of the alleles involved. Niv-44 has been a stable white for more than 70 years now. Its molecular analysis showed that another transposable element, Tam2, is integrated within exon 2 of the chs gene. This 5 kb element apparently is defective in that is does not encode a functional transposase but it has highly structured ends, the terminal 13 bp are sequence identical to Tam1, which otherwise does not show much homology. However, this observation was relevant for the analysis of the phenomenon of paramutation.

First of all it was relatively easy to show by Southern blottings that no gross alterations at the nivea locus occurred in paramutant progeny. These molecularly were perfect heterozygotes. Even though F2s did not reveal phenotypic differences, molecularly they were of the expected homozygous and heterozygous types. Because of the sequence identity at the ends of Tam1 and Tam2 we felt that something else than mutation might be going on. To test for possible physiological interactions between these two nivea alleles characterized by the integrations of Tam1 and Tam2, respectively, a mutant allele of Tam1 was used which stabilized the mobility if Tam1.

At one end of Tam1 5 bp were deleted thus preventing movement of the element. Because of this the phenotype was stably white. This allele was called niv-46. If the stable white niv-46 was crossed to the stable white niv-44 one would expect exclusively white progeny. However, surprisingly variegated flowers appeared, even on plants which mostly were white flowering variegated branches were seen. Clippings and regeneration of new flowers from variegated branches as well as selfings of the variegated flowers always resulted in white flowers. Clearly the phenomenon was transient, restricted to particular stage of development. Apparently the system escaped from paramutagenicity at these stages thus resulting in variegated flowers.

Occasionally we observed a very early event leading to a full red flower. Molecular analysis revealed that Tam2 had become excised thus restoring an active chs gene but leaving a diagnostic footprint behind.

Apparently Tam1 transposase which could not act on Tam1 ends because of the 5 bp deletion in one of them can, however, recognize the Tam2 ends, excise and hence transpose this element. So far so good, but should this transactivation not increase variegation in F1 or in subsequent progeny, and, why is transactivation transient.

An easy assumption is that niv-44 contains a suppressor inhibiting or interfering with Tam1 transposase. To test for such a negatively acting factor, a niv-44 line, in which Tam2 and part of nivea had been deleted, niv-4432, was used in a cross with niv-53. Clearly niv-4432 was still paramutagenic, hence the factor could not have been encoded by Tam2 at the nivea locus. Sequence analysis of this Tam2 had told us already that it was unlikely that this element encoded a protein. However, niv-44 contains quite a number of Tam2 elements in its genetic background some of which are transcribed into RNA and hence could encode the postulated inhibitor. If the line niv-44 contained a few genes encoding a suppressor of Tam1 transposase activity then indeed the lack of segregation of the variegated phenotypes in selfes of niv-53 x niv-44 crosses could be explained. Hence a crossing program was devised to segregate out the suppressor genes. Lines were obtained which either had one or none of the suppressor genes. If the latter is used in a cross with niv-53 all progeny is variegated and if the one copy line is used instead, half of the progeny is variegated due to heterozygosity at the suppressor locus. Whether or not this locus is identical with one of the many Tam2 elements present in this line remains to be determined molecularly.

However, we favor this assumption because in the En/Spm system of maize we have shown molecularly that a truncated En-encoded protein severely can reduce En-transposase mediated excision.

If this also would apply to the Tam1/Tam2 situation this would indicate that paramutation in this instance is not a mutational phenomenon but a physiological interaction of different but related transposable elements.

## 3.1.7 Open Questions

In a certain sense, the molecular study of maize transposable elements has been satisfying. As predicted, transposable element *Ac* is not so different from bacterial or *Drosophila* transposable elements. On the other hand, some of the phenomena described in detail by McClintock and her followers are not yet understood in biochemical terms, and new questions have emerged during the biochemical studies. I shall try to enumerate a few of them:

1. Transposition is a rare event, when compared with other biochemical reactions. A particular copy of the transposable element, inserted in a known location in the genome, transposes from this location less than once per cell. Why is this event so rare? The pattern of reversion that can be detected at the phenotypical level indicates that many cells are capable of sustaining transposition, but only a few of them experience this transposition in a given plant. Transcription of the elements is rare and for a while we thought that this reaction may be the limiting factor. Recently, recalibrations of our experiments have cast doubts on this assumption. From our own experiments and from those of others, we now have rea-

sons to believe that transcription is frequent enough to provide one or two mRNA molecules for each cell.

If it were true that transcription of *Ac* occurs in every cell, something else must be responsible for the fact that transposition occurs only in some of them. It is conceivable that the DNA molecules themselves are not always susceptible to the action of the transposase. In this case, however, transposition of one transposable element within a given cell should not be correlated to the transposition of another copy of the same element present in another location of the genome. McClintock has described experiments which make it likely that events within one cell can be correlated, and other experiments support these old observations. If this were so, another reason for the rare transposition should be looked for. A hypothetical possibility is the presence of a second protein product of *Ac*. As the known mRNA spans nearly the whole element, no unused regions offer themselves for this function. Alternative splicing, however, is not impossible and if the splice product would be very rare, as has been observed with maize transposable element *En* and *Drosophila* transposable element P, this might be the rate-limiting factor.

2. *Ac* transposition is dependent on regulatory factors that are only incompletely known. An old observation, again by McClintock, is the dose effect. If two doses of *Ac* are present within a cell, transposition occurs later during endosperm development, and the frequency of transposition events is often (but not always) decreased. As both transcription and translation of *Ac* are increased with *Ac* copy number, there seems to be a negative regulation of transposition. The observation that not only fewer transposition events occur, but that these also occur at a different time during plant development indicates an interaction with unknown cellular factors that change during plant development.

It will not be easy to understand this regulation in biochemical terms. The task is even made more difficult by the observation that the effect of an increase of *Ac* dose not always as described above. In tobacco, an increase in the *Ac* product (by an increase in dose, as was shown by Jonathan Jones in Great Britain) seems to stimulate transposition rather than to decrease it. Hypotheses have been formulated and experiments are under way to clarify this situation, but it is too early to quote any results.

3. Transposable element *Ac* occurs only in some maize lines and always in small numbers. This is not true, however, if one looks at DNA sequences related to *Ac*. From the results of our very first experiments we surprisingly learned that the number of DNA sequences related to *Ac* is higher than 50 and that these *Ac*-related sequences are found in all maize lines tested. There is an easy evolutionary explanation for this phenomenon. If plants have a selective disadvantage, if transposition becomes too frequent, the inactivation of *Ac* will confer a selective advantage to the host

plants. This hypothesis might explain a negative dosage effect. It might also explain the frequent inactivation of the transposable element by mutation. It is further supported by the observation that *Ac* can become temporarily inactive, that this inactivation is accompanied by methylation, and that in the inactive and methylated state transcription ceases.

Though this hypothesis is satisfying, it cannot be the sole explanation. Some of the mutated *Ac*-elements carry internal deletions. They might have originated, as described above. Others, however, have undergone much more pronounced alterations. A family of *Ac*-related sequences of complicated composition is described in the report by Hans-Peter Döring. However, it is even possible that completely unrelated sequences are incorporated into an inactive *Ac*-element or *Ds*-element, as they are usually called. One of the first of these was described by the laboratory of Jim Peacock, and another one was characterized in our laboratory. Most of them do not occur singly, but rather in multiple copies. They cannot have been created separately. Rather, they must have multiplied as such. In the case of one of these elements, we could even show that the family of these elements undergoes further evolution and diversifies. The reasons for this are completely unclear.

Thus, the study of maize transposable element *Ac* is another example that despite all the successes of molecular genetics in recent years, we are far from an understanding of the genetic structure of living organisms. If society chooses to support the increasingly expensive research that is necessary to solve the questions posing themselves at the present level of understanding, the number of these questions is certainly not too small to keep biologists at universities and other research institutions busy.

## 3.1.8   Publications

### 3.1.8.1   On the IS-Work

E. Jordan, H. Saedler, P. Starlinger: Strong-polar mutations in the transferase gene of the galactose operon in E. coli. Mol. Gen. Genet. 100 (1967) 296–306.

H. Saedler, P. Starlinger: O° mutations in the galactose operon in E. coli. II. Mol. Gen. Genet. 100 (1967) 190.

E. Jordan, H. Saedler, P. Starlinger: O° and strongpolar mutations in the gal operon are insertions. Mol. Gen. Genet. 102 (1968) 353–363.

G. Michaelis, H. Saedler, P. Venkov, P. Starlinger: Two insertions in the galactose operon having different sizes but homologous DNA sequences. Mol. Gen. Genet. 104 (1969) 371–377.

H. Saedler, J. Besemer, B. Kemper, B. Rosenwirth, P. Starlinger: Insertion mutations in the control region of the gal operon of E. coli. Mol. Gen. Genet. 115 (1972) 258–265.

H. J. Hirsch, H. Saedler, P. Starlinger: Insertion mutations in the control region of the galactose operon of E. coli. Mol. Gen. Genet. 115 (1972) 266–267.

H. J. Hirsch, P. Starlinger, P. Brachet: Two kinds of insertions in bacterial genes. Mol. Gen. Genet. 119 (1972) 191–206.

P. Starlinger, H. Saedler: Insertion mutations in microorganisms. Biochimie 54 (1972) 177–185.

H. Saedler, B. Heiß: Multiple copies of the insertion-DNA sequences IS1 and IS2 in the chromosome of *E. coli* K12. Mol. Gen. Genet. 122 (1973) 267.

H. Saedler, H.-J. Reif, S. Hu, N. Davidson: IS2, a genetic element for turn-off and turn-on of gene activity. Mol. Gen. Genet. 132 (1974) 265.

S. Hu, E. Ohtsubo, N. Davidson, H. Saedler: Electron microscope heteroduplex studies of sequence relations among bacterial plasmids: Identification and mapping of the insertion sequences IS1 and IS2 in F and R plasmids. J. Bacteriol. 122 (1975) 764.

H. J. Reif, H. Saedler: IS1 is involved in deletion formation in the *gal* region of *E. coli* K12. Mol. Gen. Genet. 137 (1975) 17.

H. Saedler, D. F. Kubai, M. Nomura, S. R. Jaskunas: IS1 and IS2 mutations in the ribosomal-protein genes of *E. coli* K12. Mol. Gen. Genet. 141 (1975) 185.

F. Schmidt, J. Besemer, P. Starlinger: The isolation of IS1 and IS2 DNA. Mol. Gen. Genet. 145 (1976) 145–154.

P. Starlinger, H. Saedler: IS elements in microorganisms. Curr. Topics Microbiol. Immunol. 75 (1976) 112–146.

H. Chadwell, P. Starlinger: Specificity of integration of insertion sequence (IS) elements. Microbiol. 1978 (1978): 22–24.

S. Kühn, H. J. Fritz, P. Starlinger: Close vicinity of IS1 integration sites in the leader sequence of the gal operon of E. coli. Mol. Gen. Genet. 167 (1979) 235–241.

P. Habermann, R. Klaer, S. Kühn, P. Starlinger: IS4 is found between eleven or twelve base pair duplications. Mol. Gen. Genet. 175 (1979) 369–373.

P. Starlinger: Review: IS elements and transposons. Plasmid 3 (1980) 241–259.

E. Klaer, D. Pfeifer, P. Starlinger: IS4 is still found at its chromosomal site after transposition to galT. Mol. Gen. Genet. 178 (1980) 281–284.

R. Klaer, P. Starlinger: IS4 at its chromosomal site in E. coli K12. Mol. Gen. Genet. 178 (1980) 285–291.

R. Klaer, S. Kühn, H. J. Fritz, E. Tillmann, I. Saint-Girons, P. Habermann, D. Pfeifer, P. Starlinger: Studies on transposon mechanisms and specificity of IS4. Cold Spring Harbor Symp. Quant. Biol. XLV (1981) 215–224.

R. Klaer, S. Kühn, E. Tillmann, H. J. Fritz, P. Starlinger: The sequence of IS4. Mol. Gen. Genet. 181 (1981) 169–175.

K. Trinks, P. Habermann, K. Beyreuther, P. Starlinger, R. Ehring: An IS4-encoded protein is synthesized in minicells. Mol. Gen. Genet. 182 (1981) 183–188.

I. Saint-Girons, H. J. Fritz, C. Shaw, E. Tillmann, P. Starlinger: Integration specificity of an artificial Kanamycin transposon constructed by the in vitro insertion of an internal Tn5 fragment into IS2. Mol. Gen. Genet. 183 (1981) 45–50.

P. Habermann, P. Starlinger: Bidirectional deletions associated with IS4. Mol. Gen. Genet. 185 (1982) 215–222.

## 3.1.8.2   On the Work with Plants

M. Geiser, H. P. Döring, J. Wöstemeyer, U. Behrens, E. Tillmann, P. Starlinger: A cDNA clone from *Zea mays* endosperm sucrose synthase mRNA. Nucl. Acids Res. 8 (1980) 6175–6188.

J. Wöstemeyer, U. Behrens, A. Merckelbach, M. Müller, P. Starlinger: Translation of *Zea mays* endosperm sucrose synthase mRNA *in vitro*. Eur. J. Biochem. 114 (1981) 39–44.

H. P. Döring, M. Geiser, P. Starlinger: Transposable element *Ds* at the *shrunken* locus in *Zea mays*. Mol. Gen. Gent. 184 (1981) 377–380.

M. Geiser, E. Weck, H. P. Döring, W. Werr, U. Courage-Tebbe, E. Tillmann, P. Starlinger: Genomic clones of a wild type allele and a transposable element-induced mutant allele of the sucrose synthase gene of *Zea mays* L. EMBO J. 1 (1982) 1455–1460.

U. Courage-Tebbe, H. P. Döring, N. Fedoroff, P. Starlinger: The controlling element *Ds* at the *shrunken* locus in *Zea mays*: Structure of the unstable *sh-m5933* and several revertants. Cell 34 (1983) 383–393.

H. P. Döring, E. Tillmann, P. Starlinger: DNA sequence of transposable element *Dissociation* in maize. Nature 307 (1984) 127–131.

H. P. Döring, M. Freeling, S. Hake, M. A. Johns, R. Kunze, A. Merckelbach, F. Salamini, P. Starlinger: A *Ds* mutation of the *Adh1* gene in *Zea mays* L. Mol. Gen. Genet. 193 (1984) 199–204.

U. Behrens, N. Fedoroff, A. Laird, M. Müller-Neumann, P. Starlinger, J. Yoder: Cloning of the *Zea mays* controlling element *Ac* from the *wx-m7* allele. Mol. Gen. Genet. 194 (1984) 346–347.

P. Starlinger: Transposable Elements. TIBS 9 (1984) 125–127.

E. Weck, U. Courage, H. P. Döring, N. Fedoroff, P. Starlinger: Analysis of *sh-m6233*, a mutation induced by the transposable element *Ds* in the sucrose synthase gene of *Zea mays*. EMBO J. 3 (1984) 1713–1716.

H. P. Döring, P. Starlinger: Barbara McClintock's controlling elements: Now at the DNA level. Cell 39 (1984) 253–259.

M. Müller-Neumann, J. Yoder, P. Starlinger: The DNA sequence of the transposable element *Ac* of *Zea mays* L. Mol. Gen. Genet. 198 (1984) 19–24.

U. Bonas, H. Sommer, B. J. Harrison, H. Saedler: The transposable element Tam1 of *Antirrhinum majus* is 17 kb long. Mol. Gen. Genet. 194 (1984) 138–143.

U. Bonas, H. Sommer, H. Saedler: The 17-kb Tam1 element of *Antirrhinum majus* induces a 3-bp duplication upon integration into the chalcone synthase gene. EMBO J. 3 (1984) 1015–1019.

K. Upadhyaya, H. Sommer, E. Krebbers, H. Saedler: The Paramutagenic Line niv-44 has a 5 kb insert, Tam2, in the Chalcone Synthase Gene of *Antirrhinum majus*. Mol. Gen. Genet. 199 (1985) 201–207.

W. Werr, W. B. Frommer, C. Maas, P. Starlinger: Structure of the sucrose synthase gene on chromosome 9 of *Zea mays* L. EMBO J. 4 (1985) 1373–1380.

A. Merckelbach, H. P. Döring, P. Starlinger: The abberant *Ds* element in the *adh1–2F11::Ds* allele. Maydica 31 (1986) 109–122.

H. P. Döring, P. Starlinger: Molecular genetics of transposable elements in plants. Ann. Rev. Genet. 20 (1986) 175–200.

B. Springer, W. Werr, P. Starlinger, D. C. Bennett, M. Zokolica, M. Freeling: The *Shrunken* gene on chromosome 9 of *Zea mays* L. is expressed in various plant tissues and encodes an anaerobic protein. Mol. Gen. Genet. 205 (1987) 461–468.

R. Simon, P. Starlinger: Transposable element *Ds2* of *Zea mays* influences polyadenylation and splice site selection. Mol. Gen. Genet. 209 (1987) 198–199.

R. Kunze, U. Stochaj, J. Laufs, P. Starlinger: Transcription of transposable element *Activator* (*Ac*) of *Zea mays* L. EMBO J. 6 (1987) 1555–1563.

N. Theres, T. Scheele, P. Starlinger: Cloning of the *Bz2* locus of *Zea mays* using the transposable element *Ds* as gene tag. Mol. Gen. Genet. 209 (1987) 193–197.

B. Baker, G. Coupland, N. Fedoroff, P. Starlinger, J. Schell: Phenotypic assay for excision of the maize controlling element *Ac* in tobacco. EMBO J. 6 (1987) 1547–1554.

P. Starlinger, B. Baker, B. Coupland, R. Kunze, J. Laufs, J. Schell, U. Stochaj: Studies on transposable element *Ac* of *Zea mays*. In: O. E. Nelson (ed.): Plant Transposable Elements. Plenum Press, New York 1987, pp. 91–100.

B. Baker, G. Coupland, R. Hehl, N. Fedoroff, H. Lörz, P. Czernilofsky, P. Starlinger, J. Schell: Transposition of *Ac* in tobacco. In: O. E. Nelson (ed.): Plant Transposable Elements. Plenum Press, New York 1987, pp. 161–174.

R. Hehl, H. Sommer, H. Saedler: Interaction between the Tam1 and Tam2 transposable elements of *Antirrhinum majus*. Mol. Gen. Genet. 207 (1987) 47–53.

E. Krebbers, R. Hehl, R. Piotrowiak, W.-E. Lönning, H. Sommer, H. Saedler: Molecular analysis of paramutant plants of *Antirrhinum majus* and the involvement of transposable elements. Mol. Gen. Genet. 209 (1987) 499–507.

W. B. Frommer, P. Starlinger: DNaseI hypersensitive sites in the 5'-region of the maize *Shrunken* gene in nuclei from different organs. Mol. Gen. Genet. 212 (1988) 351–359.

P. Starlinger, G. Coupland, H. Fußwinkel, M. Heinlein, R. Kunze, J. Laufs, M. Li, U. Stochaj, B. Baker, J. Schell, C. Both, J. Lee, C. Oellig, W. Doerfler, D. Schwartz: Studies on transposable element *Ac* in *Zea mays*. In: R. Goldberg (ed.): The Molecular Basis of Plant Development. UCLA Symp. on molecular and cellular biology, New Series Vol. 92. Alan R. Liss. Inc., NY 1988, pp. 65–75.

S. Knapp, G. Coupland, H. Uhrig, P. Starlinger, F. Salamini: Transposition of the maize transposable element *Ac* in *Solanum tuberosum*. Mol. Gen. Genet. 213 (1988) 285–290.

R. Kunze, P. Starlinger, D. Schwartz: DNA methylation of the maize transposable element *Ac* interferes with its transcription. Mol. Gen. Genet. 214 (1988) 325–327.

G. Coupland, B. Baker, J. Schell, P. Starlinger: Characterization of the maize transposable element *Ac* by internal deletions. EMBO J. 7 (1988) 3653–3659.

C. Hauser, H. Fußwinkel, J. Li, C. Oellig, R. Kunze, M. Müller-Neumann, M. Heinlein, P. Starlinger, W. Doerfler: Overproduction of the protein encoded by the maize transposable element *Ac* in insect cells by a baculovirus vector. Mol. Gen. Genet. 214 (1988) 373–378.

M. Heinlein, P. Starlinger: Tissue- and cellspecific expression of the two sucros synthase isoenzymes in developing maize kernels. Mol. Gen. Genet. 215 (1989) 441–446.

R. Kunze, P. Starlinger: The putative transposase of transposable element *Ac* from *Zea mays* L. interacts with subterminal sequences of *Ac*. EMBO J. 8 (1989) 3177–3185.

G. Coupland, C. Plum, S. Chatterjee, A. Post, P. Starlinger: Sequences near the termini are required for transposition of the maize transposon *Ac* in transgenic tobacco plants. Proc. Natl. Acad. Sci. USA 86 (1989) 9385–9388.

C. Lechelt, T. Peterson, A. Laird, J. Chen, S. L. Dellaporta, E. Dennis, W. J. Peacock, P. Starlinger: Isolation and molecular analysis of the maize *P* locus. Mol. Gen. Genet. 219 (1989) 225–234.

# List of Former Doctoral Students who were Involved in Work on Transposable Elements Since the Foundation of the Sonderforschungsbereich 74

Ursula Bonas (1984): In vitro-Klonierung eines transponierbaren Elementes im Gen der Chalkon-Synthase von *Antirrhinum majus.*

Sabine Burckhardt (1985): I. Ein offenes Leseraster von IS4. II. Cos-abhängige Stimulation von Red. Not working in science anymore.

Hans-Peter Döring (1982): Mutable Gene in Zea mays. Institut für Genetik, Universität Köln, FRG.

Wolf-Bernd Frommer (1987): Studien zur Chromatin-Struktur des Sucrose-Synthase-Gens in Zea mays L. Institut für Genbiologische Forschung, Berlin, FRG.

Paul Habermann (1982): IS4-abhängige bidirektionale Deletionen. Hoechst AG, Frankfurt, FRG.

Reinhard Hehl (1987): Molekulare Analyse induzierter Instabilität am *nivea* Locus von *Antirrhinum majus.*

Heinz-Josef Hirsch (1973): Elektronenmikroskopische Untersuchungen der DNA des Galaktose-Operons von Escherichia coli. RWTH Aachen, FRG.

Roland Klaer (1982): Strukturelle und funktionelle Studien des transponierbaren DNA-Elementes IS4. Universitätskliniken, Köln, FRG.

Reinhard Kunze (1987): Transkription des transponierbaren Elementes *Ac* aus *Zea mays* L. Institut für Genetik, Universität Köln, FRG.

Christa Lechelt (1988): Molekulare Analyse des P-Locus aus Zea mays L. Bayer AG, Leverkusen, FRG.

Armin Merckelbach (1986): Die *DS2*-Insertion im *Adh-2F11*-Allel. Beschreibung eines neuen Elementes der *Ac/Ds*-Transposonfamilie und zweier Revertanten des *Adh1-2F11*-Allels aus *Zea mays.* RheinBiotech, Düsseldorf, FRG.

Li Min-gang (1989): Funktionelle Ausprägung von mutierten Derivaten des transponierbaren Elementes *Ac* aus Mais. University of Minnesota, St. Paul, Minnesota, USA.

Markus Müller-Neumann (1985): Molekulare Isolierung und Charakterisierung des transponierbaren Elementes *Activator* am *Waxy*-Locus von *Zea mays* L. BASF, Ludwigshafen, FRG.

Ralf Piotrowiak (1986): Die molekulare Struktur des transponierbaren Elementes *Tam1* aus *Antirrhinum majus.*

Bodo Rak (1976): Messenger RNA konstitutiver Revertanten der Insertionsmutante galOP::IS2-308. Institut für Biologie, Universität Freiburg, FRG.

Hans Jörg Reif (1974): IS1 abhängige Deletionsentstehung in der gal-Region von *E. coli* K12.

Friedrich Schmidt (1976): Isolation und Charakterisierung von IS1- und IS2-DNA. Bundesforschungsanstalt für Landwirtschaft, Braunschweig, FRG.

Boris Springer (1989): Analysen von Protein-DNA-Wechselwirkungen am Transkriptionsstart des *Shrunken*-Gens aus *Zea mays* L. Universität Bonn, FRG.

Klaus Theres (1986): Studien über das transponierbare Element *Ds* am *Bronze* 2-Locus von *Zea mays* L. Institut für Genetik, Universität Köln, FRG.

Marlies Thiedemann (1984): Studien zur Transposition von IS4. Not working in science anymore.

Wolfgang Werr (1985): Molekularbiologische Studien am Shrunken-Locus auf Chromosom 9 von *Zea mays* L. Institut für Genetik, Universität Köln, FRG.

*Heinz Saedler and Peter Starlinger*

## List of Former "Wissenschaftliche Assistenten", Postdoctoral Fellows and Visiting Scientists

Jürgen Besemer, Sandoz Forschungsinstitut, Wien, Austria.
Howard Chadwell, presently unknown.
George Coupland, John Innes Institute, Norwich, GB.
Lothar Fecker, Friedrich-Miescher-Institut, Tübingen, FRG.
Hans-Joachim Fritz, Universität Göttingen, FRG.
Douglas Furtek, Pennsylvania State University, USA.
Robert Garber, Ohio State University, USA.
Martin Geiser, Ciba Geigy, Basel, Switzerland.
Rosa Gromkova, University of Johannisburg, South Africa.
Siegfried Kühn, Max-Planck-Institut für Zellbiologie, Heidelberg, FRG.
Alexander Kraev, Academy of Sciences of USSR Institute of Mol. Biology, Moscow.
Enno Krebbers: Plant Genetic System, Gent, Belgium.
D. Kubai-Maroni, presently unknown.
Alan Laird, State Dept. of Science, Adelaide, Australia.
Wolf-Ekkehard Lönnig, MPI für Züchtungsforschung, Köln. FRG.
Isabelle Saint-Girons, Institut Pasteur, Paris, France.
Marcel Salanoubat, Université Paris-Sud, Orsay, France.
Christine Shaw, Newcastle, GB.
Kilash C. Upadhyaya, Indian Institute of Technology, New Delhi, India.
Penchov Venkov, University of Sofia, Bulgaria.
Edward Weck, Northrup, King & Co., Stanton
Johannes Wöstemeyer, TU Berlin, FRG.
Liliana Waltschewa, University of Sofia, Bulgaria.
John Yoder, University of California, Davis, USA.

## Habilitations

Dr. Ruth Ehring (1972), Institut für Genetik, Universität Köln, FRG.
Dr. Heinz Saedler (1973), Max-Planck-Institut für Züchtungsforschung, Köln, FRG.
Dr. Jürgen Besemer (1977), Sandoz, Wien, Austria.
Dr. Börries Kemper (1981), Institut für Genetik, Universität Köln, FRG.

# 3.2 Experiments to Analyze the Maize Transposable Elements *Ac* and *Ds* at the DNA Level

Hans-Peter Döring*

## 3.2.1 Introduction

The movement of defined pieces of DNA – usually referred to as transposition – was studied extensively during the seventies and eighties (reviewed in [1]). Transposable elements were detected in most of the genetically well-characterized organisms like *E. coli, S. cerevisiae, Drosophila melanogaster, C. elegans, Antirrhinum majus.* The discovery of transposable elements was accomplished by Barbara McClintock, who described the first transposable element – the so-called *Ds*-element – in maize in the late forties [2, 3, 4]. *Ds* stands for *Dissociation* and illustrates the property of this transposable element to provide a specific site for chromosome breakage. Transposition attempts of the original *Ds*-element characterized by McClintock often led to chromosome breakage or to chromosomal rearrangement at the insertion site of the element [5]. Upon the study of the element's transposition behavior, McClintock isolated several unstable mutations at the *Sh* locus [6, 7]. Four of these *sh* alleles and several derivatives thereof were studied in my laboratory in the last years. These experiments led to the understanding of the structure of the chromosome-breaking *Ds*-element. Two hypotheses were proposed which describe the molecular events at the *Ds* sequences during chromosome breakage. The transposition of these *Ds*-elements can cause complicated chromosomal rearrangements.

* Institut für Genetik der Universität zu Köln

## 3.2.2 Isolation of the Chromosome-Breaking *Ds*-Element

*Ds*-elements cannot transpose by themselves, but require the presence of another transposable element, the *Ac*-element. Several unstable mutant alleles caused by *Ds* or *Ac* were cloned by several groups (reviewed in [8]). These studies demonstrated that most *Ds*-elements are incomplete copies of the *Ac* sequence (e. g. internal deletions). The *Ac*-element is 4.5 kb long, has inverted repeats at its termini, and produces a characteristic target site duplication of 8 bp upon insertion. The *Ds*-sequence isolated from three mutable *sh* alleles, which all had in common the property to induce chromosome breakage, consisted of two identical copies of a 2 kb *Ds*-element. Thus, the two *Ds*-elements formed a double *Ds* structure of approximately 4 kb [9]. One of the two copies was transposed into the approximate center of the other copy, as deduced from target site duplication of the interrupted *Ds*-element copy. The orientation of both *Ds* sequences was opposite with respect to each other (Fig. 3.2.1a). A derivative of the 4 kb double *Ds* structure is approximately 3 kb long (Fig. 3.2.1b). It is sequence-identical with the corresponding parts of the 4 kb double *Ds* structure, but one half *Ds*-element is missing. Thus, the 3 kb double *Ds* structure consists of one complete and one half *Ds*-element.

Since this 4 kb double *Ds* structure or derivatives of it were found in three independent *Ds*-induced *sh* alleles, it was tempting to conclude that it is this **double** structure which is responsible for chromosome breakage.

This notion was supported by several observations and findings:

1. From McClintock's genetic experiments and also from field experiments of ourselves, we know that *Ds*- or *Ac*-induced mutations now known to be caused by **single** *Ds* or *Ac* insertions are not able to induce specific chromosome breakage.

2. A certain revertant allele of the *sh-m5933* allele had undergone a deletion in the 3 kb double *Ds* structure [10, 11]. This event led to the excision of the complete 2 kb *Ds*-element, leaving behind one half *Ds*-element (Fig. 3.2.1b and 3.2.1c). This excision was accompanied by an alteration in the frequency of chromosome breakage during development of the endosperm.

3. This finding was confirmed by the isolation of nine maize strains, which displayed a similarly altered chromosome breakage pattern as the above mentioned revertant allele. Molecular analysis of these nine derivative strains revealed that similar excision events, as observed in the revertant had occurred in all nine cases [12]. 22,000 gametes were tested to find these nine independent cases.

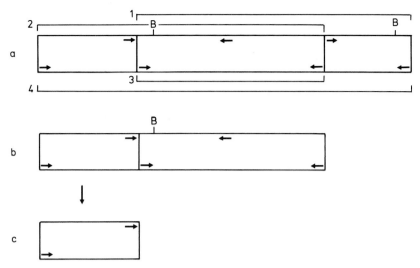

**Fig. 3.2.1:** Schematic representation of the 4 kb double *Ds* structure and derivatives. Arrows at the bottom part of the boxes indicate the inverted termini of the *Ds*-elements. Two pairs of inverted termini are present. The arrows at the top part of the box indicate the 8 bp target site duplication, which was created upon the insertion of the internal *Ds*-element into a copy of itself. The BamH I (B) is given to show the opposite orientation of the two *Ds*-elements comprising the 4 kb double *Ds* structure. (a) The complete 4 kb double *Ds* structure as it was found inserted in the *sh-m6233* allele. The numbers 1–4 illustrate the four different excision events which could be possible. Note that for 1 and 2 the end regions are inverted over a distance of approximately 1 kb, whereas for 3 and 4 only the 11 bp termini are inverted. The 4 kb double *Ds* structure is also present at one side of the 30 kb insert of the *sh-m5933* allele. On the other side, a 3 kb double *Ds* structure was found, which is schematically shown under (b). This double *Ds* structure consists of one complete and one half *Ds*-element. If the complete 2 kb *Ds*-element is excised, one half *Ds*-element is retained (c). This excision event is correlated with an alteration in the chromosome breakage pattern.

4.  In the experiment in which the nine maize strains were isolated, we also looked for maize kernels which showed no sign of chromosome breakage. These kernels were not found. This result suggests that the excision of the 3 kb double *Ds* structure is not possible [12].

5.  We also failed to detect the product of the excision of the 3 kb double *Ds* structure with the polymerase chain reaction. We concluded that the attempt to excise the 3 kb double *Ds* structure was not working properly. The result of this faulty excision was chromosome breakage at the insertion site of the double *Ds* structure (Korfhage and Döring, unpublished results).

From the five points discussed above it is concluded that double *Ds* structures are involved in specific chromosome breakage.

## 3.2.3   How to Explain Chromosome Breakage at the Double *Ds* Structure

There are two possibilities to explain chromosome breakage at the double *Ds* structure.

– The first hypothesis [13] proposes that the transposase recognizes two terminal sequences which are in the same orientation. This would be conceivable since in the 4 kb double *Ds* structure, there are two pairs of inverted termini and thus, two combinations of direct repeats of the sequence making the inverted termini (Fig. 3.2.1a). If the transposase would have cut the DNA at the first nucleotide of the terminal sequences present in the same orientation, and if sequences adjacent to these cleavage sites would be ligated, the result would be a chromosome breakage. Whereas the orientation of the duplicated *Ds*-elements is not important for this hypothesis, the orientation of the two *Ds*-elements with respect to each other is decisive for chromosome breakage in the second hypothesis.

– The second hypothesis [12] says that for a correct transposition process, the transposase need more sequence information than the short 11 bp sequence of the inverted termini. It is proposed that the *Ds* sequences beyond the termini are also needed for the interaction between the transposase and the *Ds*-element. Thus, it is anticipated that a *Ds*-element with two identical ends would cause difficulties upon excision.

For the 4 kb double *Ds* structure, different excision events can be envisioned. The excision of the complete 4 kb double *Ds* structure, and the excision of the internal 2 kb *Ds*-element should make no problems because the ends are not identical except for the inverted termini (Fig. 3.2.1a, no. 3 and 4). In opposition to these excisions, the attempted excision of one of the two 3 kb double *Ds* structures comprising one complete and one half *Ds*-element would face the problem of two identical ends of approximately 1 kb (Fig. 3.2.1, no. 1 and 2). It is proposed that 3 kb excisions are very unlikely or impossible and that they lead to chromosome breakage or other types of rearrangements.

It is possible that the transposase cannot properly recognize the 3 kb double *Ds* structure because both the terminal regions are inverted over a distance of approximately 1 kb, or alternatively, because the right and the left ends of the *Ds*-element are not present at the terminal regions of both the 3 kb double *Ds* structures. In these structures, we have either two left or two right terminal regions present at the ends. At present, it cannot be determined which of these alternatives is correct.

## 3.2.4 Chromosome-Breaking *Ds*-Elements can Generate Complicated Chromosomal Rearrangements

The pure 4 kb double *Ds* structure is the insertion which causes the unstable *sh-m6233* allele [14] (Fig. 3.2.1a). The 4 kb double *Ds* structure is part of a complicated chromosomal rearrangement in the unstable *sh-m5933* and *sh-m6258* alleles.

The *sh-m5933* allele is caused by the insertion of a DNA segment of approximately 30 kb into the seventh intron of the sucrose synthase gene [11, 12]. This segment consists of two closely related double *Ds* structures of 4 kb and 3 kb, respectively. These double *Ds* structures and neighboring sequences of several kb are bordering the insertion on both sides in inverted orientation. The remaining internal DNA segment is not part of the inverted duplication and is present only once. The structure of the 30 kb insert demonstrates that *Ds*-elements can be used as building blocks to form large insertions. Thus, *Ds*-elements are to mobilize DNA sequences unrelated to themselves.

Structurally similar chromosomal rearrangements were observed cytogenetically by McClintock [5] when she analyzed the transposition behavior of the chromosome-breaking *Ds*-element.

An additional complication of the *sh-m5933* allele is the presence of a large duplication. This duplication comprises part of the 30 kb insertion, including one of the double *Ds* structures and the adjacent sequences of the 5' region of the *Sh* locus.

In revertant derivatives of the *sh-m5933* allele the 30 kb insertion is excised, whereas the large duplication is retained [10, 11]. The unstable *sh-m6258* allele is also caused by a long insertion. The length is at least 45 kb. The inserted fragment is also bordered by *Ds* sequences. The neighboring internal sequences are different at each side (Pahl and Döring, unpublished results). A revertant derivative of the *sh-m6258* allele carries a 2 kb *Ds*-element at the same location of the 45 kb insertion. One possibility to explain the generation of the revertant allele is unequal cross-over between the distal and the proximal *Ds* sequences of the long insert of the progenitor allele.

## 3.2.5 How was the Symmetrical Insert in the *sh-m5933* Allele Created?

The 30 kb insert of the *sh-m5933* allele was formed by an faulty excision event of the 4 kb double *Ds* structure from its position in the progenitor strain. It is assumed that the 4 kb double *Ds* was recognized and cleaved by the transposase only at one terminal sequence. The second cut was introduced in a certain distance from the double *Ds* structure. If these events happened simultaneously on both chromatids, ligation of these DNA segments at the non-*Ds* cleavage sites and subsequent transpositions of this partially symmetric DNA segment into the *Sh* gene could have generated the structure of the 30 kb insert as it was found in the *sh-m5933* allele. The observation that the double *Ds* structures are slightly different on both sides of the 30 kb insert can be explained. It is conceivable that different *Ds* termini were recognized by the transposase. The 30 kb insert is bordered by the complete 4 kb double *Ds* structure on one side and by the shorter 3 kb derivative on the other side. The 30 kb insert is flanked by a target site duplication of 8 bp [12]. Thus, the 30 kb was indeed moved to its position in the *sh* gene by a transposition, rather than some other rearrangement.

## 3.2.6 References

[1] D. E. Berg, M. M. Howe (eds.): Mobile DNA. American Society for Microbiology, Washington, D. C. 1989.
[2] B. McClintock: Maize genetics. Carnegie Inst. Wash. Yearbook 45 (1946) 176–186.
[3] B. McClintock: Cytogenetic studies of maize and neurospora. Carnegie Inst. Wash. Yearbook 46 (1947) 146–152.
[4] B. McClintock: Mutable loci in maize. Carnegie Inst. Wash. Yearbook 47 (1948) 155–169.
[5] B. McClintock: Mutable loci in maize. Carnegie Inst. Wash. Yearbook 49 (1950) 157–167.
[6] B. McClintock: Mutable loci in maize. Carnegie Inst. Wash. Yearbook 51 (1952) 212–219.
[7] B. McClintock: Mutations in maize. Carnegie Inst. Wash. Yearbook 52 (1953) 227–237.
[8] H.-P. Döring, P. Starlinger: Molecular genetics of transposable elements in plants. Ann. Rev. Genet. 20 (1986) 175–200.
[9] H.-P. Döring, E. Tillmann, P. Starlinger: DNA sequence of the maize transposable element *Dissociation*. Nature 307 (1984) 127–130.

[10] U. Courage-Tebbe, H.-P. Döring, N. Fedoroff, P. Starlinger: The controlling element *Ds* at the *Shrunken* locus in *Zea mays*: Structure of the unstable *sh-m5933* allele and several revertants. Cell 34 (1983) 383–393.
[11] H.-P. Döring, R. Garber, B. Nelsen, E. Tillmann: Transposable element *Ds* and chromosomal rearrangements. In: M. Freeling (ed.): Genetics. Liss, New York, 1985, pp. 355–367.
[12] H.-P. Döring, B. Nelsen-Salz, R. Garber, E. Tillmann: Double *Ds* elements are involved in specific chromosome breakage. Mol. Gen. Genet. 219 (1989) 299–305.
[13] H.-P. Döring, P. Starlinger: Barbara McClintock's Controlling Elements: Now at The DNA Level. Cell 39 (1984) 253–259.
[14] E. Weck, U. Courage, H.-P. Döring, N. Fedoroff, P. Starlinger: Analysis of *sh-m6233*, a mutation induced by the transposable element *Ds* in the sucrose synthase gene of *Zea mays*. EMBO J. 3 (1984) 1713–1716.

## 3.2.7  Publications and Review Articles

## Publications

M. Geiser, H.-P. Döring, J. Wöstemeyer, U. Behrens, E. Tillmann, P. Starlinger: A cDNA clone from *Zea mays* endosperm sucrose synthase mRNA. Nucleic Acids Res. 8 (1980) 6175–6188.

H.-P. Döring, M. Geiser, P. Starlinger: Transposable element *Ds* at the *shrunken* locus of *Zea mays*. Mol. Gen. Genet. 184 (1981) 377–380.

M. Geiser, E. Weck, H.-P. Döring, W. Werr, U. Courage-Tebbe, E. Tillmann, P. Starlinger: Genomic clones of a wild type allele and a transposable element-induced mutant allele of the sucrose synthase gene of *Zea mays*. EMBO J. 1 (1982) 1455–1460.

U. Courage-Tebbe, H.-P. Döring, N. Fedoroff, P. Starlinger: The controlling element *Ds* at the *Shrunken* locus in *Zea mays*: Structure of the unstable *sh-m5933* allele and several revertants. Cell 34 (1983) 383–393.

H.-P. Döring, E. Tillmann, P. Starlinger: DNA sequence of transposable element *Dissociation* in maize. Nature 307 (1984) 127–131.

H.-P. Döring, M. Freeling, S. Hake, M. A. Johns, R. Kunze, A. Merckelbach, F. Salamini, P. Starlinger: A *Ds* mutation at the *Adh1* gene in *Zea mays* L. Mol. Gen. Genet. 193 (1984) 199–204.

E. Weck, U. Courage, H.-P. Döring, N. Fedoroff, P. Starlinger: Analysis of *sh-m6233*, a mutation induced by the transposable element *Ds* in the sucrose synthase gene of *Zea mays*. EMBO J. 3 (1984) 713–716.

A. Merckelbach, H.-P. Döring, P. Starlinger: The aberrant *Ds* element in the *adh1–2F11::Ds2* allele. Maydica 31 (1986) 109–122.

H.-P. Döring, B. Nelsen-Salz, R. Garber, E. Tillmann: Double *Ds* elements are involved in chromosome breakage. Mol. Gen. Genet. 219 (1989) 299–305.

*Hans-Peter Döring*

## Review Articles

H.-P. Döring, P. Starlinger: Barbara McClintock's controlling elements: Now at the DNA level. Cell 39 (1984) 253–259.

H.-P. Döring: Plant Transposable Elements. BioEssays 3 (1985) 164–171.

H.-P. Döring, P. Starlinger: Molecular Genetics of Transposable Elements in Plants. Ann. Rev. Genet. 20 (1986) 175–200.

H.-P. Döring: Tagging Genes with Transposable Elements. An Overview. Maydica 34 (1989) 73–88.

# 3.3 *Agrobacterium* – Mediated Gene Transfer to Plants: Its Impact on Plant Breeding, Studies of Gene Expression and Plant Developmental Biology

Jeff Schell*

## 3.3.1 Introduction

In September 1978, we came to the Max-Planck-Institut für Züchtungsforschung to assume responsibility for a new Department: Genetische Grundlagen der Pflanzenzüchtung (Genetic Principles of Plant Breeding). The mandate was to initiate and develop research on Plant Molecular Biology relevant to Plant Breeding at the Max-Planck-Institut für Züchtungsforschung. It was recognized that this goal could only be achieved in close collaboration with the University of Köln and in particular with the Institut für Genetik because of the leading role played by this Institute in molecular biology research in Germany and Europe. We were, therefore, very interested in the Sonderforschungsbereich 74 and were glad to be able to join it in 1978. Our nomination as "Honorarprofessor" at the "Mathematisch-Naturwissenschaftliche Fakultät der Universität zu Köln" (Feb. 1980) was an important further step in the same direction.

The primary goal of our work in the first several years after joining the Sonderforschungsbereich was the elucidation of the molecular basis of crown gall formation on plants by *Agrobacterium tumefaciens* and the mechanism of DNA transfer to higher plants. This research formed the basis of the development of effective gene vectors for plants and of methods for the introduction and stable maintenance of novel or modified genes into plants. Furthermore, the techniques and vector systems originating from the early work became essential tools in the study of gene expression in higher plants regulated in a tissue-specific manner or in response to hormones, heat shock, pathogens, light, and symbiotically nitrogen-fixing bacteria. Tissue-specific promoters were thus identified, which can be used to direct the expression of newly introduced genes in the desired tissue (organ) of the plant.

* Max-Planck-Institut für Züchtungsforschung Köln-Vogelsang

Historically, the following 'mile-stones' could be mentioned: Transposon mutagenesis of the large extrachromosomal plasmid in *Agrobacterium tumefaciens* (Ti plasmid) revealed two segments (*vir*, T-region) to be involved in transfer of bacterial DNA into the plant. It could be shown that the T-region, as well as transposons or other foreign DNA inserted into it, was physically integrated into the genome of transformed cells, thus establishing that Ti plasmids could be used to introduce foreign DNA into plant cells. Subsequent genetic experiments revealed that such DNA was transmitted in a Mendelian fashion.

As none of the genes located in the T-region were found to be involved in the mechanism responsible for T-DNA transfer and integration into the plant genome, 'disarmed vectors', no longer containing oncogenic phytohormone genes, could be constructed. In addition, binary vectors consisting of (1) a helper Ti plasmid from which the whole T-region, including border sequences, was deleted and (2) a broad-host-range plasmid flanked by the T-region border sequences in the proper polarity were devised. The use of these disarmed vectors allowed genetic engineering via *Agrobacterium* in many plants, including several of the major crop plants, with the exception of cereals.

The function of individual T-DNA genes was determined by using clonal tobacco cell lines containing a single T-DNA gene or a defined combination of T-DNA genes. For example, T-DNA genes 1, 2, and 4 were found to play a major role in the control of neoplastic growth. Cells expressing only genes 1 and 2 produced root-generating tissues. Those expressing only gene 4 grew as shoot-producing teratomas. These observations showed that the combination of genes 1 and 2 establishes a new pathway for auxin synthesis in transformed plant cells, whereas gene 4 codes for the synthesis of a cytokinin, which was confirmed by biochemical studies.

The *cis*-acting elements involved in the regulation of gene expression of T-DNA encoded loci in plants were analyzed. Genes participating in opine synthesis and phytohormone production were shown to exhibit several common features of eukaryotic promoters. These promoter elements were essential to construct chimeric genes, consisting of, for example, opine synthesis gene promoter and polyadenylation sequences flanking the coding sequences of bacterial antibiotic resistance genes. These chimeric genes were found to be properly expressed in transgenic tissues, conferring resistance to the antibiotic, and constituted the first generation of selectable marker genes for *Agrobacterium*-mediated plant transformation experiments. The ability to create transgenic plants, harboring chimeric plant promoter-reporter genes via *Agrobacterium*-mediated gene transfer, also opened the way to narrow down specific regulatory (promoter) sequences. In studies of light-mediated gene expression, for example, it could be shown that a fragment derived from a pea *rbcS* gene could drive the light- and chloroplast-dependent expression of a CAT (chloramphenicolacetyl transferase) reporter gene. The role of enhancer elements in the light- and chloroplast-dependent expression of other nuclear genes (LHCP a/b, ST-LSI) could also be demonstrated. Subsequent studies on tuber-specific (patatin) or wound-inducible (*wun*) genes in potatoes and nitrogen-fixing, nodule-specific leghemoglobin (*lb*) genes in legumes demonstrated the role of distinct 5' upstream regulatory elements in the tissue-specific expression of these genes. Distinct *cis*-acting regulatory regions in the pathogen or U. V. light-induced chalcone synthase gene of soybeans were also identified using these methods.

These *cis*-acting elements have been characterized for their interaction with sequence-specific *trans*-acting regulatory proteins involved in tissue- or stimulus-specific gene expression. It has been possible to construct vectors in which these enhancer and silencer regulatory elements are used to drive the regulated expression of foreign genes in plants. This has made a significant contribution to being able to express newly introduced genes in transgenic plants in a highly regulated manner, which is essential for applied purposes.

*Agrobacterium*-mediated transformation, leading to random integration of T-DNA segments, was used for gene tagging experiments to identify and clone plant genes of interest. Ti plasmid vectors were specifically designed for this purpose.

More recently, work on the mechanisms of signal transduction in plants was initiated. Genes encoding putative hormone (auxin) receptor proteins were cloned and analyzed. In addition, genes encoding guanine-nucleotide binding proteins (G-proteins), calcium-binding proteins, and protein kinases are now being characterized. This line of research constitutes a major part of the present and future research direction in our laboratory, namely signal transduction and plant development.

The *Agrobacterium rhizogenes*-plant interaction was also exploited to study signal transduction and plant development. Specific *A. rhizogenes* loci (*rol*A, B and C) were found to drastically alter plant development when expressed singly or in combination in transgenic plants, due to distinct effects of their gene products directly or indirectly on plant growth regulator activities.

The symbiotic *Rhizobium*-legume interaction constitutes an additional, important model system to study signal transduction. The common nodulation genes of *R. meliloti* (*nod* ABC) were shown to encode a receptor-like protein (NodC), as well as a diffusible, hormone-like factor, which is capable of inducing cell division (meristematic activity) in plant tissues (NodAB). Introduction and expression of these genes in transgenic plants were also found to severely affect plant development.

These projects serve as examples for our continued interest in Plant-Microbe interactions, first studied to elucidate plant transformation. Now, ten years later they are used to study signal transduction and development.

*Jeff Schell*

The present work in our department is organized around the major themes summarized below. The main thrust of the research is the identification and characterization of genes (gene products) involved in the control of plant growth and development. Moreover, the capacity of the laboratory will be increasingly used to carry out more fundamental research with applied goals, such as fungal and viral resistance in cereals and qualitative plus quantitative modifications of fatty acid biosynthesis in oil-seed crops. Most of the projects listed are interrelated and designed to provide basic information, techniques, vectors, and genetic material to reach these goals.

## 3.3.2 Plant Developmental Biology

Three main areas of research form the basis of this research theme:

### 3.3.2.1 Identification, Cloning, and Characterization of Plant Genes Involved in Signal Transduction

– Molecular cloning of genes from *Zea mays* coding for putative auxin receptors.
– Molecular cloning and functional analysis of genes coding for GTP-binding proteins.
– Molecular cloning and structural analysis of genes encoding calcium-binding proteins and protein kinases.

### 3.3.2.2 Elucidation of the Mechanism of Action of Oncogenic T-DNA Genes from *A. rhizogenes*

The so-called *rol* gene products were shown to induce marked growth and differentiation abnormalities in a cell-autonomous fashion. The *rol*B and *rol*C gene products probably interfere with auxin signal transduction in opposite ways, and their precise role is being analyzed.

### 3.3.2.3 Identification of Genes Involved in Control of Plant Development by Gene-Tagging Mutagenesis

– *Agrobacterium*-mediated T-DNA transfer: By selection for insertional gene fusions several interesting genes were tagged and subsequently isolated from both *Nicotiana* and *Arabidopsis*.

128

– Maize *Ac-Ds* transposable elements are being optimized as efficient gene-tags for *Arabidopsis*. Screens for the identification of homeotic genes are being carried out.

# 3.3.3 Regulation of Gene Expression in Plants

A number of different genes are under study with the aim to first identify the *cis*-acting elements and later the *trans*-acting factors involved in the regulation of their expression.

*Cis*-acting elements from these genes are being identified by the construction of chimeric genes and by the analysis of their expression in transgenic plants and in some cases (chalcone synthase) in transient expression assays.

*Trans*-acting factors (proteins) are looked for by retardation of DNA fragments on gels and by affinity chromatography using oligonucleotides, the sequence of which is derived from a combination of genetic analysis, footprinting, and binding competition experiments. The work has advanced the most with chalcone synthase and leghemoglobin genes.

Insertion mutagenesis vectors have been designed to tag genes encoding *trans*-acting factors, and attempts are being made to clone such genes by screening plant gene expression libraries with defined oligonucleotides, as well as via direct genetic selection for DNA-binding domains in *E. coli* and yeast.

## 3.3.3.1 Light Inducible Genes

– *rbcS* of Rubisco: All the members of this gene family in potatoes have been isolated and their relative expression is being monitored, using newly modified PCR methods.
– Chalcone synthase of *Antirrhinum majus* and soybean: The correlation of genetically identified *cis*-elements with binding proteins has been extended, leading to a more detailed regulation model, which is being examined.

## 3.3.3.2 Organ-Specific Genes

The major effort has been on nodule-specific genes and in particular on the leghaemoglobin genes. The interaction of defined *cis*-elements with nodule-specific binding proteins (*trans*-acting factors) is being analyzed. The role of the *cis*-acting elements is being examined, using chimeric β-glucuronidase (*gus*) genes in transgenic *Lotus* and tobacco plants.

### 3.3.3.3   Stress Induced Genes

The wound-inducible potato gene, *wun*1, is being studied in great detail. A fragment of about 1000 bp 5' upstream to the transcription initiation site of *wun*1 has been shown to be responsible for the wound-induced transcription of chimeric genes. Fully induced, the *wun*1 promoter is a strong promoter (comparable to the Cauliflower Mosaic Virus 35S promoter). The function of *wun*1 is still unknown but it appears to be involved in either suberinisation or in callose formation.

## 3.3.4   Transformation of Cereal Plants

The most promising approach appears to be the uptake of DNA in cells of dessicated embryos. This method has been shown to lead to efficient transient expression. By combining this method of DNA-uptake with the use of a replicating genome of the Wheat Dwarf Virus (WDV), stable and inheritable transformation of cereals will hopefully be achieved.

## 3.3.5   The Rhizobium-Legume Symbiosis as a Model System for Plant-Bacterial Interactions

The (regulatory) signals from plant and bacterial origin, which are essential for the establishment of a nitrogen-fixing nodule, are being investigated. Three symbiotic systems have been chosen for our analysis: *Rhizobium meliloti* – Alfalfa, *Bradyrhizobium japonicum* – soybean, and the unusual stem-nodulation *Azorhizobium caulinodans* ORS571 – *Sesbania rostrata* system. The following topics are being addressed:

– Rhizobial nitrogen fixation (*nif*), assimilation (*asm*) and regulation (*ntr*) genes, and the role of oxygen, combined nitrogen and potential plant derived symbiotic control signals in *nif/asm* gene regulation.
– The products of rhizobial "early nodulation" (*nod*) genes and their function.

The *nod*C gene product has been shown to be a transmembrane protein with a receptor-like structure. The *nod*AB gene products have been found to be involved in the synthesis of a diffusable factor, which is capable of

130

stimulating cell division in soybean protoplasts. The nature (structure) of this factor (compound) is being determined. The implications of these findings for plant-microbe signal transduction pathways (see also 3.3.2) are being investigated.

## 3.3.6 Molecular Biology and Potential Application of DNA Plant Viruses

Wheat dwarf virus (WDV), a geminivirus, infects a wide variety of graminaceous monocots, including most of the cereals. The single-stranded genome consists of one circular DNA molecule of 2.75 kb. Virus transmission from plant to plant exclusively occurs via a cicadellid leafhopper.

The molecular biology of WDV is studied with regard to (I) the life cycle of the virus and (II) its suitability to independently replicate and express foreign genes in monocotyledonous plants.

The lack of mechanic transmission of the virus of its DNA to plants demands the development of an assay system to follow the replication of the viral genome in tissue culture. This has been established by transfecting protoplasts of different monocot suspension culture cells with the various cloned WDV DNAs. For a period of up to three weeks post transfection, the replication of the viral genome can be followed, and the replicative intermediates in culture cells are the same as in naturally infected plant tissue.

This system has been used to define the WDV genes and regions that are essential or dispensable for replication. In particular, the capsid protein gene has been replaced by a variety of marker genes, whose expression was monitored. The genes used include the neomycin phosphotransferase gene (*npt*II), the chloramphenicol acetyltransferase gene (*cat*), the β-galactosidase gene (*lacZ*), and a cDNA of the presumed transposase of the element *Ac*. All these genes are replicated and expressed.

Currently recombinant WDV-vectors are being used to analyze the movement of the virus in plants after mechanical treatment with DNA, following agroinfection and embryo imbibition.

## 3.3.7 Projects with a Long Term Applied Goal

1. Cloning of Barley yellow mosaic virus and Barley mild mosaic virus.
2. Expression of antifreeze proteins from winter flounder in tobacco.
3. Functional expression of *Eco*RI-*Eco*MI restriction-modification system in tobacco.
4. Bacterial luciferase as a reporter gene.
5. Improvement of oil content in oil-seed crops.

## 3.3.8 Publications

M. Claeys, E. Messens, M. van Montagu, J. Schell: GC/MS determination of cytokinins in *Agrobacterium tumefaciens* cultures. Fresenius Z. Anal. Chem. 290 (1978) 125–126.

A. Depicker, M. van Montagu, J. Schell: A DNA region, common to all Ti-plamids, is essential for oncogenicity. Arch. Int. Physiol. Biochim. 86 (1978) 422–423.

A. Depicker, M. van Montagu, J. Schell: Homologous DNA sequences in different Ti-plasmids are essential for oncogenicity. Nature 275 (1978) 150–153.

M. de Wilde, M. Depicker, G. de Vos, M. de Beuleleer, E. van Haute, M. van Montagu, J. Schell: Molecular cloning as a tool to the analysis of the Ti-plasmids of *A. tumefaciens.* Ann. Microbiol. (Inst. Pasteur) 129 B (1978) 531–532.

J. P. Hernalsteens, H. de Greve, M. van Montagu, J. Schell: Mutagenesis by insertion of the drug resistance transposon Tn7 applied to the Ti-plasmid of *Agrobacterium tumefaciens.* Plasmid 1 (1978) 218–225.

J. P. Hernalsteens, H. de Greve, M. van Montagu, J. Schell: A technique for mutagenesis by transposon insertion, applicable to most gram-negative bacteria. Arch. Int. Physiol. Biochim. 86 (1978) 432–433.

M. Holsters, D. de Waele, A. Depicker, E. Messens, M. van Montagu, J. Schell: Transfection and transformation of *Agrobacterium tumefaciens.* Mol. Gen. Genet. 163 (1978) 181–187.

M. Holsters, A. Silva, C. Genetelleo, G. Engler, F. van Vliet, M. de Block, R. Villarroel, M. van Montagu, J. Schell: Spontaneous formation of cointegrates on the oncogenic Ti-plasmid and the wide-host-range P-plasmid RP. Plasmid 1 (1978) 456–467.

M. Holsters, A. Silva, F. van Vliet, J. P. Hernalsteens, C. Genetello, M. van Montagu, J. Schell: *In vivo* transfer of the Ti-plasmid of *Agrobacterium tumefaciens* to *Escherichia coli.* Mol. Gen. Genet. 163 (1978) 335–336.

A. Petit, J. Tempe, A. Kerr, M. Holsters, M. van Montagu, J. Schell: Substrate induction of conjugative activity of *Agrobacterium tumefaciens* Ti plasmids. Nature 271 (1978) 570–572.

J. Schell: On the transfer and expression of bacterial plasmid DNA in higher plants. Arch. Int. Physiol. Biochim. 86 (1978) 901–902.

F. van Vliet, A. Silva, M. van Montagu, J. Schell: Transfer of RP4:: Mu plasmids to *Agrobacterium tumefaciens.* Plasmid 1 (1978) 446–455.

M. H. Whatley, J. B. Margot, J. Schell, B. B. Lippincott, J. A. Lippincott: Plasmid and chromosomal determination of *Agrobacterium* adherence specifity. J. Gen. Microbiol. 107 (1978) 395–398.

M. Aerts, M. Jacobs, J. P. Hernalsteens, M. van Montagu, J. Schell: Induction and *in vitro* culture of *Arabidopsis thaliana* crown gall tumors. Plant. Sci. Lett. 17 (1979) 43–50.

P. Dhaese, H. de Greve, H. Decraemer, J. Schell, M. van Montagu: Rapid mapping of transposon insertion and deletion mutations in the large Ti-plasmids of *Agrobacterium tumefaciens*. Nucl. Acids Res. 7 (1979) 1837–1849.

J. G. Ellis, A. Kerr, M. van Montagu, J. Schell: *Agrobacterium*: genetic studies on agrocin 84 production and the biological control of crown gall. Physiol. Plant Pathol. 15 (1979) 311–319.

J. Schell, M. van Montagu, M. de Beuckeleer, M. de Block, A. Depicker, A. de Wilde, G. Engler, C. Genetello, J. P. Hernalsteens, M. Holsters, J. Seurinck, A. Silva, F. van Vliet, R. Villarroel: Interactions and DNA transfer between *Agrobacterium tumefaciens*, the Ti-plasmid and the plant host. Proc. R. Soc. Lond. B 204 (1979) 251–266.

L. Debrouwere, M. van Montagu, J. Schell: The *ral* gene of phage lambda. III. Interference with *E. coli* ATP dependent funcitons. Mol. Gen. Genet. 179 (1980) 81–88.

L. Debrouwere, M. Zabeau, M. van Montagu, J. Schell: The *ral* gene of phage lambda. II. Isolation and characterization of *ral* deficient mutants. Mol. Gen. Genet. 179 (1980) 75–80.

A. Depicker, M. de Block, D. Inze, M. van Montagu, J. Schell: IS-like element IS8 in RP4 plasmid and its involvement in cointegration. Gene 10 (1980) 329–338.

A. Depicker, M. de Wilde, G. de Vos, R. de Vos, M. van Montagu, J. Schell: Molecular cloning of overlapping segments of the nopaline Ti-plasmid pTiC58 as a means to restriction endonuclease mapping. Plasmid 3 (1980) 193–211.

J. P. Hernalsteens, F. van Vliet, M. de Beuckeleer, A. Depicker, G. Engler, M. Lemmers, M. Holsters, M. van Montagu, J. Schell: The *Agrobacterium tumefaciens* Ti plasmid as a host vector system for introducing foreign DNA in plant cells. Nature 287 (1980) 654–656.

P. R. Hirsch, M. van Montagu, A.W. B. Johnston, N. J. Brewin, J. Schell: Physical identification of bacteriocinogenic, nodulation and other plasmids in strains of *Rhizobium leguminosarum*. J. Gen. Microbiol. 120 (1980) 403–412.

M. Holsters, B. Silva, F. van Vliet, C. Genetello, M. de Block, P. Dhaese, A. Depicker, D. Inze, G. Engler, R. Villarroel, M. van Montagu, J. Schell: The functional organization of the nopaline *A. tumefaciens* plasmid pTiC58. Plasmid 3 (1980) 212–230.

J. Leemans, R. Villarroel, R. Maenhaut, M. van Montagu, J. Schell: Degrees of homology between some P-Type R plasmids. Arch. Int. Physiol. Biochim. 88 (1980) B38–B39.

J. Leemans, R. Villarroel, B. Silva, M. van Montagu, J. Schell: Direct repitition of a 1.2 Md DNA sequence is involved in site-specific recombination by the P1 plasmid R 68. Gene 10 (1980) 319–328.

M. Lemmers, M. de Beukeleer, M. Holsters, P. Zambryski, A. Depicker, J. P. Hernalsteens, M. van Montagu, J. Schell: Internal organization, boundaries and integration of Ti-plasmid DNA in nopaline crown gall tumours. J. Mol. Biol. 144 (1980) 355–378.

M. van Montagu, M. Holsters, P. Zambryski, J. P. Hernalsteens, A. Depicker, M. de Beuckeleer, G. Engler, M. Lemmers, L. Willmitzer, J. Schell: The interaction of *Agrobacterium* Ti-plasmid DNA and plant cells. Proc. R. Soc. Lond. B210 (1980) 351–365.

L. Willmitzer, M. de Beuckeleer, M. Lemmers, M. van Montagu, J. Schell: DNA from Ti plasmid present in nucleus and absent from plastids of crown gall. Nature 287 (1980) 359–361.

M. Zabeau, S. Friedman, M. van Montagu, J. Schell: The *ral* gene of phage lambda. I. Identification of a non-essential gene that modulates restriction and modification in *E. coli*. Mol. Gen. Genet. 179 (1980) 63–73.

P. Zambryski, M. Holsters, K. Kruger, A. Depicker, J. Schell, M. van Montagu, H. M. Goodman: Tumor DNA structure in plant cells transformed by *A. tumefaciens*. Science 209 (1980) 1385–1391.

M. de Beuckeleer, M. Lemmers, G. de Vos, L. Willmitzer, M. van Montagu, J. Schell: Further insight on the transferred-DNA of octopine crown gall. Mol. Gen. Genet. 183 (1981) 283–288.

H. de Greve, H. Decraemer, J. Seurinck, M. van Montagu, J. Schell: The functional organization of the octopine *Agrobacterium tumefaciens* plasmid pTiB6DS3. Plasmid 6 (1981) 235–248.

G. de Vos, M. de Beuckeleer, M. van Montagu, J. Schell: Restriction endonuclease mapping of the octopine tumor inducing plasmid pTiAch5 of *Agrobacterium tumefaciens*. Plasmid 6 (1981) 249–253.

P. Dhaese, M. van Montagu, J. Schell: Transcription of nopaline T-DNA sequences in a tobacco crown gall teratoma. Arch. Int. Physiol. Biochim. 89 (1981) B163–B164.

G. Engler, A. Depicker, R. Maenhaut, R. Villarroel, M. van Montagu, J. Schell: Physical mapping of DNA base sequence homologies between an octopine and a nopaline Ti-plasmid of *Agrobacterium tumefaciens*. J. Mol. Biol. 152 (1981) 183–208.

J. Leemans, Ch. Shaw, R. Deblaere. H. de Greve, J. P. Hernalsteens, M. Maes, M. van Montagu, J. Schell: Site-specific mutagenesis of *Agrobacterium* Ti plasmids and transfer of genes to plant cells. J. Mol. Appl. Genet. 1 (1981) 149–164.

M. Lemmers, M. Holsters, G. Engler, M. van Montagu, J. Leemans, H. de Greve, J. P. Hernalsteens, L. Willmitzer, L. Otten, J. Schröder, J. Schell: Le plasmide Ti, vecteur potentiel pour la modification genetique des plantes. C.R. Acad. Agric. France 67, no. 12 (1981) 1052–1065.

L. Otten, H. de Greve, J. P. Hernalsteens, M. van Montagu, O. Schieder, J. Straub, J. Schell: Mendelian transmission of genes introduced into plants by the Ti plasmids of *Agrobacterium tumefaciens*. Mol. Gen. Genet. 183 (1981) 209–213.

J. Schröder, G. Schröder, H. Huisman, R. A. Schilperoort, J. Schell: The mRNA for lysopine dehydrogenase in plant tumor cells is complementary to a Ti-plasmid fragment. FEBS Lett. 129 (1981) 166–168.

L. Willmitzer, L. Otten, G. Simons, W. Schmalenbach, J. Schröder, G. Schröder, M. van Montagu, G. de Vos, J. Schell: Nuclear and polysomal transcripts of T-DNA in octopine crown gall suspension and callus cultures. Mol. Gen. Genet. 182 (1981) 255–262.

L. Willmitzer, W. Schmalenbach, J. Schell: Transcription of T-DNA in octopine and nopaline crown-gall tumours is inhibited by low concentrations of alpha-amanitin. Nucl. Acids Res. 9 (1981) 4801–4812.

H. de Greve, J. Leemans, J. P. Hernalsteens, L. Thia-Toong, M. de Beuckeleer, L. Willmitzer, L. Otten, M. van Montagu, J. Schell: Regeneration of normal and fertile plants that express octopine synthase, from tobacco crown galls after deletion of tumour-controlling functions. Nature 300 (1982) 752–755.

M. Holsters, R. Villarroel, M. van Montagu, J. Schell: The use of selectable markers for the isolation of plant-DNA/T-DNA junction fragments in a cosmid vector. Mol. Gen. Genet. 185 (1982) 283–289.

J. Leemans, R. Deblaere, L. Willmitzer, H. de Greve, J. P. Hernalsteens, M. van Montagu, J. Schell: Genetic identification of functions of TL-DNA transcripts in octopine crown galls. EMBO J. 1 (1982) 147–152.

J. Leemans, J. Langenakens, H. de Greve, R. Deblaere, M. van Montagu, J. Schell: Broad-host-range cloning vectors derived from the W-plasmid Sa. Gene 19 (1982) 361–364.

L. Willmitzer, J. Sanchez-Serrano, E. Buschfeld, J. Schell: DNA from *Agrobacterium rhizogenes* is transferred to and expressed in axenic hairy root plant tissues. Mol. Gen. Genet. 186 (1982) 16–22.

L. Willmitzer, G. Simons, J. Schell: The TL-DNA in octopine crown gall tumours code for seven well defined polyadenylated transcripts. EMBO J. 1 (1982) 139–146.

A. Caplan, L. Herrera-Estrella, D. Inze, E. van Haute, M. van Montagu, J. Schell, P. Zambryski: Introduction of genetic material into plant cells. Science 222 (1983) 815–821.

H. de Greve, P. Dhaese, J. Seurinck, M. Lemmers, M. van Montagu, J. Schell: Nucleotide sequence and transcript map of the *Agrobacterium tumefaciens* Ti plasmid-encoded octopine synthase gene. J. Mol. Appl. Genet. 6 (1983) 499–511.

P. Dhaese, H. de Greve, J. Gielen, J. Seurinck, M. Montagu, J. Schell: Identification of sequences involved in the polyadenylation of higher plant nuclear transcripts using *Agrobacterium* T-DNA genes as model. EMBO J. 2 (1983) 419–426.

M. Holsters, R. Villarroel, J. Gielen, J. Seurinck, H. de Greve, M. van Montagu, J. Schell: An analysis of the boundaries of the octopine TL-DNA in tumors induced by *Agrobacterium tumefaciens*. Mol. Gen. Genet. 190 (1983) 35–41.

L. Herrera-Estrella, A. Depicker, M. van Montagu, J. Schell: Expression of chimaeric genes transferred into plant cells using a Ti plasmid-derived vector. Nature 303 (1983) 209–213.

L. Herrera-Estrella, M. de Block, E. Messens, J. P. Hernalsteens, M. van Montagu, J. Schell: Chimeric genes as dominant selectable markers in plant cells. EMBO J. 2 (1983) 987–995.

D. Inze, E. van Haute, M. van Montagu, J. Schell: Localization of the ornithine-catabolism genes (s) (*orc*) on the *Agrobacterium tumefaciens* plasmid pTiC58. Arch. Int. Physiol. Biochim. 91 (1983) B105–B106.

H. Joos, D. Inze, A. Caplan, M. Sormann, M. van Montagu, J. Schell: Genetic analysis of T-DNA transcripts in nopaline crown galls. Cell 32 (1983) 1057–1067.

H. Joos, B. Timmermann, M. van Montagu, J. Schell: Genetic analysis of transfer and stabilization of *Agrobacterium* DNA in plant cells. EMBO J. 2 (1983) 2151–2160.

C. Koncz, H. de Greve, D. Andre, F. Deboeck, M. van Montagu, J. Schell: The octopine synthase genes carried by Ti plasmids contain all signals necessary for expression in plants. EMBO J. 2 (1983) 1597–1603.

J. Schell, M. van Montagu: The Ti plasmids as natural and as practical gene vectors for plants. Bio/Technology 1 (1983) 175–180.

C. H. Shaw, J. Leemans, C. H. Shaw, M. van Montagu, J. Schell: A general method for the transfer of cloned genes to plant cells. Gene 23 (1983) 315–330.

E. van Haute, H. Joos, M. Maes, G. Warren, M. van Montagu, J. Schell: Intergenic transfer and exchange recombination of restriction fragments cloned on pBR322: a novel strategy for the reversed genetics of the Ti plasmids of *Agrobacterium tumefaciens*. EMBO J. 2 (1983) 411–418.

R. Villarroel, R. W. Hedges, R. Maenhaut, J. Leemans, G. Engler, M. van Montagu, J. Schell: Heteroduplex analysis of P-plasmid evolution: the role of insertion and deletion of transposable elements. Mol. Gen. Genet. 189 (1983) 390–399.

L. Willmitzer, P. Dhaese, P. Schreier, W. Schmalenbach, M. van Montagu, J. Schell: Size, localization and polarity of T-DNA-encoded transcripts in nopaline crown gall tumors; evidence for common transcripts present in both octopine and nopaline tumors. Cell 32 (1983) 1045–1056.

P. Zambryski, H. Joos, C. Genetello, J. Leemans, M. van Montagu, J. Schell: Ti plasmid vector for the introduction of DNA into plant cells without alteration of their normal regeneration capacity. EMBO J. 2 (1983) 2143–2150.

M. de Block, L. Herrera-Estrella, M. van Montagu, J. Schell, P. Zambryski: Expression of foreign genes in regenerated plants and in their progeny. EMBO J. 3 (1984) 1681–1689.

J. Gielen, M. de Beuckeleer, J. Seurinck, F. Deboeck, H. de Greve, M. Lemmers, M. van Montagu, J. Schell: The complete nucleotide sequence of the TL-DNA of the *Agrobacterium tumefaciens* plasmid pTiAch5. EMBO J. 3 (1984) 835–846.

R. Hain, H.-H. Steinbiss, J. Schell: Fusion of *Acrobacterium* and *E. coli* spheroplasts with *Nicotiana tabacum* protoplasts – direct gene transfer from microorganisms to higher plants. Plant Cell Reports 3 (1984) 60–64.

J. P. Hernalsteens, L. Thia-Toong, J. Schell, M. van Montagu: An *Agrobacterium*-transformed cell culture from the monocot Asparagus officinalis. EMBO J. 3 (1984) 3039–3041.

L. Herrera-Estrella, G. van den Broeck, R. Maenhaut, M. van Montagu, J. Schell: Light-inducible and chloroplast-associated expression of a chimaeric gene introduced into *Nicotiana tabacum* using a Ti plasmid vector. Nature 310 (1984) 115–120.

D. Inze, A. Follin, M. van Lijsebettens, C. Simons, C. Genetello, M. van Montagu, J. Schell: Genetic analysis of the individual T-DNA genes of *Agrobacterium tumefaciens*: further evidence that two genes are involved in indole-3-acetic acid synthesis. Mol. Gen. Genet. 194 (1984) 265–274.

C. Koncz, F. Kreuzaler, Z. Kalman, J. Schell: A simple method to transfer, integrate and study expression of foreign genes, such as chicken ovalbumin and alpha-actin in plant tumors. EMBO J. 3 (1984) 1029–1037.

L. Otten, H. de Greve, J. Leemans, R. Hain, P. Hooykaas, J. Schell: Restoration of virulence of *Vir* region mutants of *Agrobacterium tumefaciens* strain B6S3 by coinfection with normal and mutant *Agrobacterium* strains. Mol. Gen. Genet. 195 (1984) 159–163.

F. Salomon, R. Deblaere, J. Leemans, J. P. Hernalsteens, M. van Montagu, J. Schell: Genetic identification of functions of TR-DNA transcripts in octopine crown galls. EMBO J. 3 (1984) 141–146.

J. Schmidt, M. John, E. Kondorosi, A. Kondorosi, U. Wieneke. G. Schröder, J. Schröder, J. Schell: Mapping of protein coding regions of *Rhizobium meliloti*. EMBO J. 3 (1984) 1705–1711.

J. Velten, L. Velten, R. Hain, J. Schell: Isolation of a dual promotor fragment from the Ti plasmid of *Agrobacterium tumefaciens*. EMBO J. 3 (1984) 2723–2730.

A. Wöstemeyer, L. Otten, J. Schell: Sexual transmission of T-DNA in abnormal tobacco regenerants transformed by octopine and nopaline strains of *Agrobacterium tumefaciens*. Mol. Gen. Genet. 194 (1984) 500–507.

C. Bachem, E. Kondorosi, Zs. Banfalvi, B. Horvarth, A. Kondorosi, J. Schell: Identification and cloning of nodulation genes from a wide host range *Rhizobium* strain MPIK3030. Mol. Gen. Genet. 199 (1985) 271–278.

A. Caplan, M. van Montagu, J. Schell: Genetic analysis of integration mediated by single T-DNA borders. J. Bacteriol. 2 (1985) 655–664.

A. Cashmore, L. Szabo, M. Timko, A. Kausch, G. van den Broeck, P. Schreier, H. Bohnert, L. Herrera-Estrella, M. van Montagu, J. Schell: Import of polypeptides into chloroplasts. BioTechnology 3 (1985) 803–808.

R. Deblaere, B. Bytebier, H. de Greve, F. Deboeck, J. Schell, M. van Montagu, J. Leemans: Efficient octopine Ti-plasmid-derived vectors for *Agrobacterium*-mediated gene transfer in plants. Nucl. Acids Res. 13 (1985) 4777–4788.

M. de Block, J. Schell, M. van Montagu: Chloroplast transformation by *Agrobacterium tumefaciens*. EMBO J. 4 (1985) 1367–1372.

A. Depicker, L. Herman, A. Jacobs, J. Schell, M. van Montagu: Frequencies of simultaneous transformation with different T-DNAs and their relevance to the *Agrobacterium*/plant cell interaction. Mol. Gen. Genet. 201 (1985) 477–484.

P. Eckes, J. Schell, L. Willmitzer: Organ-specific expression of three leaf/stem specific cDNAs from potato is regulated by light and correlated with chloroplast development. Mol. Gen. Genet. 199 (1985) 216–224.

A. Follin, D. Inze, F. Budar, C. Genetello, M. van Montagu, J. Schell: Genetic evidence that the tryptophan 2-mono-oxygenase gene of *Pseudomonas savastonoi* is functionally equivalent to one of the T-DNA genes involved in plant tumour formation by *Agrobacterium tumefaciens*. Mol. Gen. Genet. 201 (1985) 178–185.

R. Hain, P. Stabel, A. P. Czernilofsky, H. H. Steinbiss, L. Herrera-Estrella, J. Schell: Uptake, integration, expression and genetic transmission of a selectable chimaeric gene by plant protoplasts. Mol. Gen. Genet. 199 (1985) 161–168.

M. John, J. Schmidt, U. Wieneke, E. Kondorosi, A. Kondorosi, J. Schell: Expression of the nodulation gene *nodC* of *Rhizobium meliloti* in *Escherichia coli*: Role of *nodC* product of nodulation. EMBO J. 4 (1985) 2425–2430.

L. Otten, G. Piotrowiak, P. Hooykaas, M. Dubois, E. Szegedi, J. Schell: Identification of an *Agrobacterium tumefaciens* pTiB6S3 vir region fragment which enhances the virulence of pTiC58. Mol. Gen. Genet. 199 (1985) 189–193.

P. H. Schreier, E. A. Seftor, J. Schell, H. J. Bohnert: The use of nuclear encoded sequences to direct the light-regulated synthesis and transport of a foreign protein into plant chloroplasts. EMBO J. 4 (1985) 25–32.

J. Simpson, M. P. Timko, A. R. Cashmore, J. Schell, M. van Montagu, L. Herrera-Estrella: Light-inducible and tissue-specific expression of a chimaeric gene under control of the 5'-flanking sequence of a pea chlorophyll a/b-binding protein gene. EMBO J. 4 (1985) 2723–2729.

A. Spena, R. Hain, U. Ziervogel, H. Saedler, J. Schell: Construction of a heat-inducible gene for plants. Demonstration of heat-inducible activity of the Drosophila hsp70 promoter in plants. EMBO J. 4 (1985) 2739–2743.

M. P. Timko, A. P. Kausch, C. Castresana, J. Fassler, L. Herrera-Estrella, G. van den Broeck, M. van Montagu, J. Schell, A. R. Cashmore: Light regulation of plant gene expression by an upstream enhancer-like element. Nature 318 (1985) 579–582.

H. van Onckelen, P. Rüdelsheim, D. Inze, A. Follin, E. Messens, S. Horemans, J. Schell, M. van Montagu, J. de Greef: Tobacco plants transformed with the *Agrobacterium* T-DNA gene *1* contain high amounts of indole-3-acetamide. FEBS Lett. 181 (1985) 373–376.

J. Velten, J. Schell: Selection-expression plasmid vectors for use in genetic transformation of higher plants. Nucl. Acids Res. 13 (1985) 6981–6998.

D. Andre, D. Colau, J. Schell, M. van Montagu, J. P. Hernalsteens: Gene tagging in plants by a T-DNA insertion mutagen that generates APH(3')II-plant gene fusions. Mol. Gen. Genet. 204 (1986) 512–518.

C. W. B. Bachem, Z. Banfalvi, E. Kondorosi, J. Schell, A. Kondorosi: Identification of host range determinants in the Rhizobium species MPIK3030. Mol. Gen. Genet. 203 (1986) 42–48.

B. Baker, J. Schell, H. Lörz, N. Fedoroff: Transposition of the maize controlling element *Activator* in tabacco. Proc. Natl. Acad. Sci. USA 83 (1986) 4844–4848.

A. P. Czernilofsky, R. Hain, L. Herrera-Estrella, H. Lörz, E. Goyvaerts, B. Baker, J. Schell: Fate of selectable marker DNA integrated into the genome of *Nicotiana tabacum.* DNA 5 (1986) 101–113.

P. Eckes, S. Rosahl, J. Schell, L. Willmitzer: Isolation and characterization of a light-inducible, organ-specific gene from potato and analysis of its expression after tagging and transfer into tobacco and potato shoots. Mol. Gen. Genet. 205 (1986) 14–22.

B. Horvarth, E. Kondorosi, M. John, J. Schmidt, I. Török, Z. Györgypal, I. Barabas, U. Wieneke, J. Schell, A. Kondorosi: Organization, structure and symbiotic function of *Rhizobium meliloti* nodulation genes determining host-specifity for alfalfa. Cell 46 (1986) 335–343.

J. S. Jensen, K. A. Marcker, L. Otten, J. Schell: Nodule-specific expression of a chimeric soybean leghemoglobin gene in transgenic *Lotus corniculatus.* Nature 321 (1986) 669–674.

H. Kaulen, J. Schell, F. Kreuzaler: Light induced expression of the chimeric chalcone synthase-NPTII gene in tobacco cells. EMBO J. 5 (1986) 1–8.

M. Keil, J. Sanchez-Serrano, J. Schell, L. Willmitzer: Primary structure of a proteinase inhibitor II gene from potato (*Solanum tuberosum*). Nucl. Acids Res. 14 (1986) 5641–5650.

C. Koncz, J. Schell: The promoter of $T_L$-DNA gene 5 controls the tissue specific expression of chimaeric genes carried by a novel type of *Agrobacterium* binary vector. Mol. Gen. Genet. 204 (1986) 383–396.

M. Kuntz, A. Simons, J. Schell, P. H. Schreier: Targeting of protein to chloroplasts in transgenic tobacco by fusion to mutated transit peptide. Mol. Gen. Genet. 205 (1986) 454–460.

S. Rosahl, P. Eckes, J. Schell, L. Willmitzer: Organ-specific gene expression in potato: isolation and characterization of tuber-specific cDNA sequences. Mol. Gen. Genet. 202 (1986) 368–373.

S. Rosahl, R. Schmidt, J. Schell, L. Willmitzer: Isolation and characterization of a gene from *Solanum tuberosum* encoding patatin, the major storage protein of potato tubers. Mol. Gen. Genet. 203 (1986) 214–220.

J. Sanchez-Serrano, R. Schmidt, J. Schell, L. Willmitzer: Nucleotide sequence of proteinase inhibitor II encoding cDNA of potato (*Solanum tuberosum*) and its mode of expression. Mol. Gen. Genet. 203 (1986) 15–20.

J. Schmidt, M. John, U. Wieneke, H.-D. Krüssmann, J. Schell: Expression of the nodulation gene *nod A* in *Rhizobium meliloti* and localization of the gene product in the cytosol. Proc. Natl. Acad. Sci. USA 83 (1986) 9581–9585.

P. H. Schreier, J. Schell: Use of chimeric genes harbouring small subunit transit peptide sequences to study transport in chloroplasts. Phil. Trans. R. Soc. Lond. B 313 (1986) 429–432.

C. Simoens, Th. Alliote, R. Mendel, A. Müller, J. Schiemann, M. van Lijsebettens, J. Schell. M. van Montagu, D. Inze: A binary vector for transferring genomic libraries in plants. Nucl. Acids Res. 14 (1986) 8073–8090.

J. Simpson, J. Schell, M. van Montagu, L. Herrera-Estrella: Light-inducible and tissue-specific pea *lhcp* gene expression involves an upstream element combining enhancer- and silencer-like properties. Nature 323 (1986) 551–554.

M. van Lijsebettens, D. Inze, J. Schell, M. van Montagu: Transformed cell clones as a tool to study T-DNA integration mediated by *Agrobacterium tumefaciens*. J. Mol. Biol. 188 (1986) 129–145.

H. van Onckelen, E. Prinsen, D. Inze, P. Rüdelsheim, M. van Lijsebettens. A. Follin, J. Schell, M. van Montagu, J. de Greef: *Agrobacterium* T-DNA gene *1* codes for tryptophan-2-monooxygenase activity in tobacco crown gall cells. FEBS Lett. 198 (1986) 357–360.

B. Baker, G. Coupland, N. Fedoroff, P. Starlinger, J. Schell: Phenotypic assay for excision of the maize controlling element Ac in tobacco. EMBO J. 6 (1987) 1547–1554.

A. De la Pena, H. Lörz, J. Schell: Transgenic rye plants obtained by injecting DNA into young floral tillers. Nature 325 (1987) 274–276.

T. Etzold, C. C. Fritz, J. Schell, P. H. Schreier: A point mutation in the chloroplast 16 S rRNA gene of a streptomycin resistant *Nicotiana tabacum*. FEBS Letters 219 (1987) 343–346.

M. J. Guiltinan, J. Velten, M. M. Bustos, R. J. Cyr, J. Schell, D. E. Fosket: The expression of a chimeric soybean beta-tubulin gene in tobacco. Mol. Gen. Genet. 207 (1987) 328–334.

U. Hilgert, J. Schell, F. J. de Bruijn: Isolation and characterization of Tn5-induced NADPH-glutamate synthase (GOGAT⁻) mutants of *Azorhizobium sesbaniae* ORS571 and cloning of the corresponding *glt* locus. Mol. Gen. Genet. 210 (1987) 195–202.

B. Horvarth, C. W. B. Bachem, J. Schell, A. Kondorosi: Host specific regulation of nodulation genes in *Rhizobium* is mediated by a plant signal, interacting with the *nodD* gene product. EMBO J. 6 (1987) 841–848.

D. Inze, A. Follin, J. Velten, L. Velten, E. Prinsen, P. Rüdelsheim, H. van Onckelen, J. Schell, M. van Montagu: The *Pseudomonas savastanoi* tryptophan-2-mono-oxygenase is biologically active in *Nicotiana tabacum*. Planta 172 (1987) 555–562.

C. Koncz, Zs. Koncz-Kalman, J. Schell: Transposon Tn5 mediated gene transfer into plants. Mol. Gen. Genet. 207 (1987) 99–105.

C. Koncz, O. Olsson, W. H. R. Langdridge, J. Schell, A. A. Szalay: Expression and assembly of functional bacterial luciferase in plants. Proc. Natl. Acad. Sci. USA 84 (1987) 131–135.

J. Logemann, J. Schell, L. Willmitzer: Improved methods for the isolation of RNA from plant tissues. Analyt. Biochem. 163 (1987) 16–20.

J. E. Mayer, G. Hahne, K. Palme, J. Schell: A simple and general plant tissue extraction procedure for two-dimensional gel electrophoresis. Plant Cell Reports 6 (1987) 77–81.

P. J. Murphy, N. Heycke, Z. Banfalvi, M. E. Tate, F. de Bruijn, A. Kondorosi, J. Tempe, J. Schell: Genes for the catabolism and synthesis of an opine-like compound in *Rhizobium meliloti* are closely linked and on the *sym* plasmid. Proc. Natl. Acad. Sci. USA 84 (1987) 493–497.

K. Pawlowski, P. Ratet, J. Schell, F. J. de Bruijn: Cloning and characterization of *nifA* and *ntrC* genes of the stem nodulating bacterium ORS571, the nitrogen fixing symbiont of *Sesbania rostrata:* Regulation of nitrogen fixation (*nif*) genes in the free living versus symbiotic state. Mol. Gen. Genet. 206 (1987) 207–219.

S. Rosahl, J. Schell, L. Willmitzer: Expression of a tuber specific storage protein in transgenic tobacco plants: Demonstration of an esterase activity. EMBO J. 6 (1987) 1155–1159.

S. Rossbach, J. Schell, F. J. de Bruijn: The *ntrC* gene of *Agrobacterium tumefaciens* C58 controls glutamine synthetase (GSII) activity, growth on nitrate and chromosomal but not Ti-encoded arginine catabolism pathways. Mol. Gen. Genet. 209 (1987) 419–426.

*Jeff Schell*

J. J. Sanchez-Serrano, M. Keil, A. O'Connor, J. Schell, L. Willmitzer: Wound-induced expression of a potato proteinase inhibitor II gene in transgenic tobacco plants. EMBO J. 6 (1987) 303–306.

J. J. Sanchez-Serrano, M. Keil, J. Schell, L. Willmitzer: Physical and functional characterization of a gene from potato encoding proteinase inhibitor II. In: Molecular Strategies for Crop Protection. Liss Inc., New York 1987, pp. 393–400.

J. St. Schell: Transgenic plants as tools to study the molecular organization of plant genes. Science 237 (1987) 1176–1183.

A. Spena, J. Schell: The expression of a heat inducible chimeric gene in transgenic tobacco plants. Mol. Gen. Genet. 206 (1987) 436–440.

A. Spena, T. Schmülling, C. Koncz, J. Schell: Independent and synergistic activity of *rol A, B* and *C* loci in stimulating abnormal growth in plants. EMBO J. 6 (1987) 3891–3899.

J. Stockhaus, P. Eckes, A. Blau, J. Schell, L. Willmitzer: Organ specific and dosage-dependent expression of a leaf/stem specific gene from potato after tagging and transfer into potato and tobacco plants. Nucl. Acids Res. 15 (1987) 3479–3491.

J. Stockhaus, P. Eckes, M. Rocha-Sosa, J. Schell, L. Willmitzer: Analysis of cis-active sequence involved in the leaf-specific expression of a potato gene in transgenic plants. Proc. Natl. Acad. Sci. USA 84 (1987) 7943–7947.

R. Töpfer, V. Matzeit, B. Gronenborn, J. Schell, H.-H. Steinbiss: A set of plant expression vectors for transcriptional and translational fusions. Nucl. Acids Res. 15 (1987) 5890.

M. Vlachova, B. A. Metz, J. Schell, F. J. de Bruijn: The tropical legume *Sesbania rostrata*: Tissue culture, plant regeneration and infection with *Agrobacterium tumefaciens* and *rhizogenes* strains. Plant Science 50 (1987) 213–223.

U. Wirtz, J. Schell, A. P. Czernilofsky: Recombination of selectable marker DNA in *Nicotiana tabacum.* DNA 6 (1987) 245–253.

A. Y. Cheung, L. Bogorad, M. van Montagu, J. Schell: Relocating a gene for herbicide tolerance: A chloroplast gene is converted into a nuclear gene. Proc. Natl. Acad. Sci. USA 85 (1988) 391–395.

G. Coupland, B. Baker, J. Schell, P. Starlinger: Characterization of the maize transposable element *Ac* by internal deletions. EMBO J. 7 (1988) 3653–3659.

A. Cseplö, T. Etzold, J. Schell, P. H. Schreier: Point mutations in the 23S rRNA genes of four lincomycin resistant *Nicotiana plumbaginifolia* mutants could provide new selectable markers for chloroplast transformation. Mol. Gen. Genet. 214 (1988) 295–299.

E. O. Jensen, K. A. Marcker, J. Schell, F. J. de Bruijn: Interaction of a nodule specific, *trans*-acting factor with distinct DNA elements in the soybean leghaemoglobin *lbc₃* 5' upstream region. EMBO J. 7 (1988) 1265–1271.

M. John, J. Schmidt, U. Wieneke, H.-H. Krüssmann, J. Schell: Transmembrane orientation and receptor-like structure of the *Rhizobium meliloti* common nodulation protein NodC. EMBO J. 7 (1988) 583–588.

S. Lipphardt, R. Brettschneider, F. Kreuzaler, J. Schell, J. L. Dangl: UV-inducible transient expression in parsley protoplasts identifies regulatory cis-elements of a chimeric *Antirrhinum majus* chalcone synthase gene. EMBO J. 7 (1988) 4027–4033.

J. Logemann, J. E. Mayer, J. Schell, L. Willmitzer: Differential expression of genes in potato tubers after wounding. Proc. Natl. Acad. Sci. USA 85 (1988) 1136–1140.

B. A. Metz, P. Welters, H.-J. Hoffmann, E. O. Jensen, J. Schell, F. J. de Bruijn: Primary structure and promoter analysis of leghemoglobin genes of the stem nodulated tropical legume *Sesbania rostrata*: Conserved coding sequences, *cis*-elements and *trans*-acting factors. Mol. Gen. Genet. 214 (1988) 181–191.

P. J. Murphy, N. Heycke, S. P. Trenz, R. Ratet, F. J. de Bruijn, J. Schell: Synthesis of an opine-like compound, a rhizopine, in alfalfa nodules is symbiotically regulated. Proc. Natl. Acad. Sci. USA 85 (1988) 9133–9137.

M. Pröls, R. Töpfer, J. Schell, H. H. Steinbiss: Transient gene expression in tobacco protoplasts: I. Time course of CAT appearance. Plant Cell Rep. 7 (1988) 221–224.

P. Ratet, K. Pawlowski, H. Meyer z. A., J. Schell, F. J. de Bruijn: Regulation of nitrogen fixation (*nif*) genes of *Azorhizobium caulinodans* ORS571 in culture and *in planta*. J. Plant Physiol. 132 (1988) 405–411.

P. Ratet, J. Schell, F. J. de Bruijn: Mini-Mu*lac* transposons with broad-host-range origins of conjugal transfer and replication for gene regulation studies in *Rhizobiaceae*. Gene 63 (1988) 41–52.

S. Rossbach. J. Schell, F. J. de Bruijn: Cloning and analysis of *Agrobacterium tumefaciens* C58 loci involved in glutamine biosynthesis: Neither the *glnA* (GSI) nor the *glnII* (GSII) gene plays a special role in virulence. Mol. Gen. Genet. 212 (1988) 38–47.

J. Schmidt, R. Wingender, M. John, U. Wieneke, J. Schell: *Rhizobium meliloti nodA* and *nodB* genes are involved in generating compounds that stimulate mitosis of plant cells. Proc. Natl. Acad. Sci. USA 85 (1988) 8578–8582.

T. Schmülling, J. Schell, A. Spena: Single genes from *Agrobacterium rhizogenes* influence plant development. EMBO J. 7 (1988) 2621–2629.

R. Töpfer, M. Pröls, J. Schell, H.-H. Steinbiss: Transient gene expression in tobacco protoplasts: II. Comparison of the reporter gene systems for CAT, NPTII, and GUS. Plant Cell Rep. 7 (1988) 225–228.

R. Töpfer, J. Schell, H.-H. Steinbiss: Versatile cloning vectors for transient expression and direct gene transfer in plant cells. Nucl. Acids Res. 16 (1988) 8725.

K. Weising, J. Schell, G. Kahl: Foreign genes in plants: transfer, structure, expression and applications. Annu. Rev. Genet. 22 (1988) 421–477.

F. P. Wolter, C. C. Fritz, L. Willmitzer, J. Schell, P. H. Schreier: *rbcS* genes in *Solanum tuberosum*: Conservation of transit peptide and exon shuffling during evolution. Proc. Natl. Acad. Sci. 85 (1988) 846–850.

F. J. de Bruijn, G. Felix, B. Grunenberg, H. J. Hoffmann, B. Metz, P. Ratet, A. Simons-Schreier, L. Szabados, P. Welters, J. Schell: Regulation of plant genes specifically induced in nitrogen-fixing nodules: role of *cis*-acting elements and *trans*-acting factors in leghemoglobin gene expression. Plant Mol. Biol. 13 (1989) 319–325.

F. J. de Bruijn, S. Rossbach, M. Schneider, P. Ratet, S. Messmer, W. W. Szeto, F. M. Ausubel, J. Schell: *Rhizobium meliloti* 1021 has three differentially regulated loci involved in glutamine biosynthesis, none of which is essential for symbiotic nitrogen fixation. J. Bacteriol. 171 (1989) 1673–1682.

I. Dusha, A. Bakos, A. Kondorosi, F. J. de Bruijn, J. Schell: The *Rhizobium meliloti* early nodulation genes (*nodABC*) are nitrogen-regulated: Isolation of a mutant strain with efficient nodulation capacity on alfalfa in the presence of ammonium. Mol. Gen. Genet. 219 (1989) 89–96.

T. Hesse, J. Feldwisch, D. Balshüsemann, G. Bauw, M. Puype, J. Vandekerckhove, M. Löbler, D. Klämbt, J. Schell, K. Palme: Molecular cloning and structural analysis of a gene from *Zea mays* (L.) coding for a putative receptor for the plant hormone auxin. EMBO J. 8 (1989) 2453–2461.

M. Kamman, J. Laufs, J. Schell, B. Gronenborn: Rapid insertional mutagenesis of DNA by polymerase chain reaction (PCR). Nucl. Acids Res. 17 (1989) 5404.

*Jeff Schell*

M. Köster-Töpfer, W. B. Frommer, M. Rocha-Sosa, S. Rosahl, J. Schell, L. Willmitzer: A class II patatin promoter is under developmental control in both transgenic potato and tobacco plants. Mol. Gen. Genet. 219 (1989) 390–396.

C. Koncz, N. Martini, R. Mayerhofer, Z. Koncz-Kalman, H. Körber, G. P. Redei, J. Schell: High frequency T-DNA mediated gene tagging in plants. Proc. Natl. Acad. Sci. USA 86 (1989) 8467–8471.

E. Kondorosi, J. Gyuris, J. Schmidt, M. John, E. Duda, B. Hoffmann, J. Schell, A. Kondorosi: Positive and negative control of *nod* gene expression in *Rhizobium meliloti* is required for optimal nodulation. EMBO J. 8 (1989) 1331–1340.

W. H. R. Langridge, K. J. Fitzgerald, C. Koncz, J. Schell, A. A. Szalay: Dual promoter of *Agrobacterium tumefaciens* mannopine synthase genes is regulated by plant growth hormones. Proc. Natl. Acad. Sci. USA 86 (1989) 3219– 3223.

J. Logemann, S. Lipphardt, H. Lörz, I. Häuser, L. Willmitzer, J. Schell: 5' upstream sequences from the *wun1* gene are responsible for gene activation by wounding in transgenic plants. The Plant Cell 1 (1989) 151–158.

J. Logemann, J. Schell: Nucleotide sequence and regulated expression of a wound-inducible potato gene (wun1). Mol. Gen. Genet. 219 (1989) 81–88.

R. V. Masterson, D. B. Furtek, C. Grevelding, J. Schell: A maize *Ds* transposable element containing a dihydrofolate reductase gene transposes in *Nicotiana tabacum* and *Arabidopsis thaliana*. Mol. Gen. Genet. 219 (1989) 461–466.

O. Olson, A. Escher, G. Sandberg, J. Schell, C. Koncz, A. A. Szalay: Engineering of monomeric bacerial luciferase by fusion of *lux*A and *lux*B genes in *Vibrio harveyi*. Gene 81 (1989) 335–347.

P. Ratet, K. Pawlowski, J. Schell, F. J. de Bruijn: The *Azorhizobium caulinodans* nitrogen fixation regulatory gene, *nifA*, is controlled by the cellular nitrogen and oxygen status. Mol. Microbiol. 3 (1989) 825–838.

B. Reiss, C. C. Wasmann, J. Schell, H. J. Bohnert: Effect of mutations on the binding and translocation functions of a chloroplast transit peptide. Proc. Natl. Acad. Sci. USA 86 (1989) 886–890.

M. Rocha-Sosa, U. Sonnewald, W. Frommer, M. Stratmann, J. Schell, L. Willmitzer: Both developmental and metabolic signals activate the promoter of a class I patatin gene. EMBO J. 8 (1989) 23–29.

H.-J. Schalk, V. Matzeit, B. Schiller, J. Schell, B. Gronenborn: Wheat dwarf virus, a geminivirus of graminaceous plants needs splicing for replication. EMBO J. 8 (1989) 359–364.

T. Schmülling, S. Beinsberger, J. de Greef, J. Schell, H. van Onckelen, A. Spena: Construction of a heat-inducible chimaeric gene to increase cytokinin content in transgenic plant tissue. FEBS Letters 249 (1989) 401–406.

T. Schmülling, J. Schell, A. Spena: Promoters of the *rolA*, *B* and *C* genes of *Agrobacterium rhizogenes* are differentially regulated in transgenic plants. The Plant Cell 1 (1989) 665–670.

B. Siebertz, J. Logemann, L. Willmitzer, J. Schell: *Cis*-analysis of the wound inducible promoter *wun1* in transgenic tobacco plants and histochemical localization of its expression. The Plant Cell 1 (1989) 961–968.

K. Spanier, J. Schell, P. H. Schreier: A functional analysis of T-DNA gene 6b: The fine tuning of cytokinin effects on shoot development. Mol. Gen. Genet. 219 (1989) 209–216.

D. Staiger, H. Kaulen, J. Schell: A CACGTG motif of the *Antirrhinum majus* chalcone synthase promoter is recognized by an evolutionary conserved nuclear protein. Proc. Natl. Acad. Sci. USA 86 (1989) 6930–6934.

142

J. Stockhaus, J. Schell, L. Willmitzer: Identification of enhancer elements in the upstream region of the nuclear photosynthetic gene ST-LS1. The Plant Cell 1 (1989) 805–813.

R. Töpfer, B. Gronenborn, J. Schell, H.-H. Steinbiss: Uptake and transient expression of chimeric genes in seed-derived embryos. The Plant Cell 1 (1989) 133–139.

P. Welters, B. A. Metz, J. Schell, F. J. de Bruijn: Nucleotide sequence of the *Sesbania rostrata* leghemoglobin (*Srglb3*) gene. Nucl. Acids Res 17 (1989) 1253.

D. Wing, C. Koncz, J. Schell: Conserved function in *Nicotiana tabacum* of a single *Drosophila hsp70* promoter heat shock element when fused to a minimal T-DNA promoter. Mol. Gen. Genet. 219 (1989) 9–16.

R. Wingender, H. Röhrig, C. Höricke, D. Wing, J. Schell: Differential regulation of soybean chalcone synthase genes in plant defence, symbiosis and upon environmental stimuli. Mol. Gen. Genet. 218 (1989) 315–322.

P. Zambryski, J. Tempe, J. Schell: Transfer and function of T-DNA genes from Agrobacterium Ti and Ri plasmids in plants. Cell 56 (1989) 193–201.

# 3.4 Resistance Mechanisms in Plants

Klaus Hahlbrock*

## 3.4.1 Summary

As a member that neither requested nor received financial support from the Sonderforschungsbereich 74, I am reporting on major aims and approaches of our work by giving the following, rather general outline.

We investigate the biochemical mechanisms and molecular genetics of disease resistance and resistance to UV irradiation in plants. Particular emphasis is placed on cytological aspects, including the temporal and spatial patterns of gene activation in the affected tissue.

Three selected systems are used for most of these studies:

- Parsley (*Petroselinum crispum*) plants and cell suspension cultures for investigations of (3.4.2) nonhost resistance to *Phytophthora megasperma* f. sp. *glycinea* (Pmg), a soybean pathogenic fungus, and (3.4.3) resistance to UV irradiation.
- Potato (*Solanum tuberosum*) plants and cell suspension cultures to study host resistance and susceptibility towards the fungal and bacterial pathogens, *Phytophthora infestans* (Pi) and *Erwinia carotovora* (3.4.4).
- Barley (*Hordeum vulgare*) for the analysis of recognition processes and the race-specific interaction with *Rhynchosporium secalis* (3.4.5).
- More recently, for several reasons *Arabidopsis thaliana* has become of interest and may be used more extensively in the future (3.4.6).

The results obtained in the last two to three years are briefly summarized here to give a general overview of the individual projects.

---

*Max-Planck-Institut für Züchtungsforschung Köln-Vogelsang

## 3.4.2 Nonhost Disease Resistance in Parsley

Cultured parsley cells and protoplasts were used to identify and characterize elicitor-active molecules from Pmg, to elucidate endogenous signal mechanisms, to identify elicitor-responsive genes, to determine their structural features, and to analyze their modes of activation. Of particular interest are putative elicitor receptors, signal transduction mechanisms, cis-acting promoter elements, and trans-acting factors regulating gene expression, as well as the genes encoding these factors.

Probes generated in these studies were used for *in situ* RNA hybridization at infection sites in Pmg-inoculated leaves. We found that the same genes were transiently activated around fungal penetration sites and in elicitor-treated cells, and the relative timing of differential gene activation for various pathways was identical. A new aspect of this work is the developmentally regulated and cell-specific expression of individual genes or members of gene families, especially where all members of metabolically related gene families are known and accessible through gene-specific probes (phenylalanine ammonia-lyase, PAL, and 4-coumarate:CoA ligase, 4CL).

## 3.4.3 UV Resistance in Parsley

Again, cultured cells and protoplasts were used for most of these studies. The major focus of this project was the mode of action of blue and UV light, and the nature and organization of cis-acting elements and their interplay with trans-acting factors. *In vivo* footprinting, in combination with *in vitro* mutation and gene-transfer experiments, has revealed a complex array of such elements on the single-copy chalcone synthase gene, as well as on selected members of the small gene families encoding PAL and 4CL. The latter are of special interest because of their involvement in both UV and disease resistance. In connection with a search for trans-acting factors, we are attempting to establish a suitable assay system for transcription initiation *in vitro*.

*In situ* RNA hybridization demonstrated the confinement of UV light-dependent chalcone synthase gene activation and flavonoid end-product accumulation to the epidermal cell layer.

## 3.4.4 Host Disease Resistance in Potato

The host resistance response of potato to Pi proved to be similar in principle to the non-host resistance response of parsley to Pmg. Probes mostly obtained through work with cultured potato cells were used for *in situ* localization of transcripts transiently accumulating in response to Pi infection of potato leaves. The following order of events was observed:
a) Very rapid, highly localized hypersensitive cell death, associated with callose deposition and accumulation of phenolic compounds in cell walls, at the site of fungal penetration; b) rapid, localized and transient activation of several defense-related pathways around infection sites (e. g., PAL, 4CL, PR1); c) relatively slow, systemic activation of genes encoding intra- as well as extracellularly accumulating PR proteins (e. g., 1,3-β-glucanase) throughout infected leaves and, to a lesser extent, in neighboring leaves.

Similar or identical reactions occurred in compatible interactions (host susceptible), but apparently more or less retarded and spreading through the tissue in a more diffuse manner. By contrast, the non-host response to Pmg was even more rapid than the resistance response to Pi. Work with *Erwinia carotovara* has just been initiated. One major aim of all these studies is the elucidation of early recognition mechanisms and the function of race-cultivar specific resistance genes.

## 3.4.5 Interaction of Barley with *Rhynchosporium*

Several toxic proteins were isolated from different races of *Rhynchosporium secalis* and tested on selected barley cultivars. This work, in combination with cytological studies, is aimed at identifying target sites for fungal effector molecules in the plant plasma membrane.

## 3.4.6 *Arabidopsis*

Studies have just been initiated to exploit the small genome size and other advantages of this system for mutant analysis, as well as the generation of probes for cytological investigations.

# 4 Development and Differentiation

# 4.1    Neurogenesis in Drosophila

José Antonio Campos-Ortega*

## 4.1.1    Introduction

One of the major problems of developmental biology is the origin of cell diversity. We approach this problem by trying to understand the mechanisms of commitment of the neural progenitor cells, the so-called neuroblasts, in *Drosophila melanogaster*. During early embryogenesis, the 2000 cells that constitute the neurogenic region of the ectoderm of the fruitfly develop into one of two different fates: 500 cells develop as neuroblasts, and the remaining 1500 cells develop as epidermal progenitor cells, or epidermoblasts. The separation of the neural from the epidermal cell lineage occurs in that individual cells from within the neuroectoderm drop into the interior of the embryo to become committed to the neural fate and build up the primordium of the central nervous system, whereas the remaining neuroectodermal cells continue their development at the outside of the embryo committed as epidermoblasts to give rise to the ventral and cephalic regions of the epidermal sheath. Thus, neighboring cells have to decide between two alternative developmental fates, the neural and the epidermal. We are trying to understand how this process is regulated.

Experiments performed in Köln have shown that the decision of a given *Drosophila* neuroectodermal cell for one of the two lineages very much depends on interactions that the cell performs with other cells in the neighborhood. The main effort of our work is to analyze the cellular and genetic mechanisms underlying the process of lineage separation. The investigated problem is highly complex and, although there is a great deal of information already available, its study is still in a fairly preliminary stage. During the last years, we succeeded in identifying the genes whose products mediate the cellular interactions required for the cells to take on one fate or the other. We have studied the functional interrelationships between the genes, cloned the

---

\* Institut für Entwicklungsphysiologie der Universität zu Köln

DNA of some of them, and established the sequence of the putative gene products. Below, I shall consider various aspects of the work done in Köln on this problem during the years 1984–1988. I shall refer first to the cellular aspects and deal subsequently with the genetic and molecular bases of neurogenesis.

## 4.1.2    Cellular Aspects of Neurogenesis

In the early *Drosophila* embryo, during embryogenetic stage 8 (staging according to [1]), the ectoderm becomes subdivided into a lateral part of small cylindrical cells and a medial one of large cuboidal cells. The lateral part differentiates during later stages into the tracheal placodes and the dorsal epidermis, whereas the medial sector is the neuroectoderm itself, from which the primordia for the ventral cord and ventral epidermis originates (Refs. [1–3]). The segregation of the neuroblasts from the neuroectoderm lasts for approximately three hours and is discontinuous, proceeding in three discrete pulses which give rise to three subpopulations, SI, SII, and SIII, of neuroblasts. Conspicuous shape changes which take place in the neuroectodermal cells during segregation, as well as the striking relationships between the various parts of the neuroectoderm and the mitotic activity in the ectodermal layer, are discussed in Hartenstein and Campos-Ortega [1, 2] and will not be discussed here any further. However, it should be mentioned that some of the SII neuroblasts, as well as most or all SIII neuroblasts, share common lineages with epidermoblasts (see [3]).

The proportion of epidermoblasts to neuroblasts is fairly reproducible from animal to animal. Roughly 25 % (approximately 500) of all cells of the neuroectoderm in the blastoderm stage develop as neuroblasts, whereas the remaining 75 % (approximately 1500 cells) develop as epidermoblasts [1, 2]. There is no evidence as to the mechanisms that restrict the production of one or the other type of cells. The timing of segregation could be one of the possible constraints that restrict the final number of each one of the progenitor cell types.

### 4.1.2.1    The Proliferation of the Neuroblasts

Immediately after segregating from the ectodermal germ layer, each neuroblast rounds up and starts dividing within 10–20 min. Within the ventral nerve cord, insect neuroblasts divide asymmetrically, following a stem cell pattern of proliferation, to give rise, after each mitosis, to a ganglion mother cell and a neuroblast that divides again asymmetrically (refer to [4]). In the

150

*Drosophila* embryo, the duration of the cycle of the ganglion mother cells was found to last for about approximately 100 min. Neuroblasts carry out up to eight cycles of division, which appear to be parasynchronous and occur at regular intervals of 40–50 min each. Since segregation takes place in three pulses that cover a time interval of 140 min, and embryonic mitotic activity of the neuroblasts is apparently interrupted at about the same time for all of them, not all of the neuroblasts divide the same number of times. Hence, during embryogenesis SI neuroblasts apparently perform 8–9, SII 6–7, and SIII 5–6 mitoses.

## 4.1.2.2   Cell Commitment in the Neurogenic Ectoderm

In insects, the decision of the neuroectodermal cells to adopt the epidermal or the neural fate is mediated by cell-cell interactions. Two pieces of experimental evidence support this conclusion. On the one hand, laser ablation experiments carried out in grasshoppers showed that the cells remaining in the neuroectoderm after the neuroblasts have segregated are not firmly commited to their fate [5]. Under normal circumstances, these cells would develop as epidermoblasts; however, in the conditions of the experiment, they may adopt the neural fate instead. These results led to propose that the prospective neuroblasts and epidermoblasts interact, such that the latter are inhibited by the former from adopting the neural fate.

On the other hand, results of cell transplantations in *Drosophila* indicate that positive regulatory signals pass between the cells of the neuroectoderm causing their commitment to either the epidermal or the neural fate. This is particularly supported by a series of heterotopic and heterochronic cell transplantations performed with both wild-type and mutant cells (Refs. [6–8]); discussed in [9]. Upon their heterotopic transplantation, dorsal ectodermal cells transplanted ventrally develop according to their new location, giving rise to either epidermal or neural clones. Since dorsal cells do not develop as neuroblasts, either *in situ* under normal conditions or upon homotopic transplantation, this observation has been interpreted to mean that the transplanted dorsal cells are actively induced by their neighbors in the neuroectoderm to adopt a neural fate. Since no intercellular influences that would actively prevent neurogenesis can be experimentally demonstrated in the dorsal region, the results support the existence of a neuralizing signal.

Further support for the hypothesis that cellular interactions mediate the segregation of lineages derives from heterochronic transplantations of ectodermal cells [8]. When ectodermal cells of increasingly older donors are transplanted into young hosts, the transplanted cells behave in the same way as younger cells in isochronic transplants and may change their fate. Heterochronic transplantations involved both epidermoblasts, i. e., cells of the dorsal epidermal anlage, and neuroblasts, i. e., cells of the neuroectoderm of so-called neurogenic mutants (see below), which as far as we know develop like wild-type neuroblasts. The neuroblasts were aged for up to 170 min after gastrulation and had, therefore, divided up to three or four times, respectively. The frequent switch of fate of the cells upon transplantation that we observed in these experiments suggests that, under normal circumstances, the segregation of the two types of progenitor cells does not by itself imply their irreversible commitment to the neural or the epidermal fate, for this fate can be changed after experimental manipulation.

# 4.1.3    Genetics of Neurogenesis

The available evidence strongly suggests that the proteins encoded by two groups of genes provide the molecular basis for the regulatory signals, which control the process of neurogenesis. The so-called *neurogenic* genes, together with a second set of various other genes, including the genes of the *achaete-scute* complex (AS-C), *ventral nervous system condensation defective* (*vnd*) and *daughterless* (*da*), are required for a proper segregation of neural and epidermal lineages during development of the central and peripheral nervous systems. The two groups of genes, considered in a global manner, apparently exert opposite effects on the differentiation of epidermal and neural lineages [9, 10].

## 4.1.3.1   The Neurogenic Genes

The complete loss of a neurogenic gene function causes the diversion of all the cells of the neuroectoderm to the neural fate. Hence, approximately 2000 cells initiate neurogenesis in the neurogenic mutants, instead of only 500 in the wild-type. This leads to embryonic lethality, associated with massive hyperplasia of the CNS and an increase in the number of sensory neurones, with concomitant lack of the entire ventrolateral and cephalic epidermis in the mature embryo (different aspects of the complex phenotype of the neurogenic mutants are described in [11, 12–17]).

### 4.1.3.1.1  Mosaic Analysis of Neurogenic Mutants

Different techniques have been used to obtain mosaics of neurogenic mutations. Homozygosity for neurogenic mutations has been induced by means of mitotic recombination throughout larval development in cells of various imaginal discs [16]. The main conclusion of that study was that in addition to their participation in the process of segregation of the neuroblasts from the epidermoblasts, the functions of the neurogenic genes are also required for normal development of the imaginal epidermal cells. An additional important conclusion to be drawn from these observations is that neurogenic gene products, with the exception perhaps of *bib,* are unable to diffuse over long distances. Moreover, the results are compatible with a cell-autonomous expression of the neurogenic genes [16], at least in the imaginal discs. Hoppe and Greenspan ([18] and personal communication) studied gynandromorph embryos formed by $N^+$ and $N^-$ cells and came to the same conclusion with respect to $N$ in its embryonic function.

However, the results of transplanting homozygous mutant ectodermal cells in the neuroectoderm of the wild-type [7] contradict the conclusion of a cell-autonomous expression of the neurogenic genes. When individual cells from the neuroectoderm of wild-type donors are transplanted homotopically and isochronically into wild-type hosts, the cells give rise to clonal progenies belonging to three different types; neural, epidermal, and mixed, i. e., some cells of the clone differentiate neural and other cells differentiate epidermal histotypes. The results of performing the same experiment using homozygous neurogenic mutant embryos as donors are, with the remarkable exception of $E(spl)^-$ cells, the same as when the donors are wild-type, i. e., neural epidermal, and mixed clones in similar proportions to those of the controls. Since the same neurogenic mutant cells would have invariably adopted neural fate while developing *in situ,* the neurogenic genes under discussion do not behave as cells autonomous in their expression, at least under the conditions of the transplantation experiment. Since single cells lacking either of the neurogenic genes *N, amx, bib, mam, neu* or *D1,* develop normally when surrounded by wild-type cells, the mutant cells appear to be capable of receiving and processing the epidermalizing signal with the same efficacy as the wild-type cells and hence, adopt the epidermal fate in some cases. These results strongly suggest that the corresponding mutant cells have normal receptor mechanisms but an abnormal signal source. In striking contrast, only neural clones develop upon transplantation of cells lacking the *E(spl)* locus. Therefore, the *E(spl)* is the only neurogenic locus with cell autonomous expression under the conditions of the transplantation experiment; moreover, $E(spl)^-$ cells cannot react to the epidermalizing signal. Thus, the *E(spl)* locus is a good candidate to encode protein(s) related to the different steps from the receptor to the nucleus, e. g., receptor molecules themselves, "second messengers" or transcription factors.

### 4.1.3.1.2 The Genetics and Molecular Genetics of the *Delta* Locus

Mutations in the *D1* gene (located in the band 92A2 of the third chromosome [19]) express a large variety of phenotypic traits pointing to considerable genetic complexity. *D1* has haplo-insufficient expression [20, 21], thus heterozygotes with a deletion of the locus, or any other amorphic *Dl* mutation, show abnormalities of the wings, compound eyes, and bristles. *Dl* deletion homozygotes are lethal at the embryonic stage and develop a severe neurogenic phenotype. Three recessive visible *Dl* alleles ($Dl^{via1}$, $Dl^{via2}$, $Dl^{via3}$) have been recovered [19], which in homozygosity cause slight delta-like thickenings, rough compound eyes, as well as a shortening and fusion of tarsal segments. Lethal *Dl* alleles exhibit a complex pattern of heteroallelic complementation [19, 22], compatible with the notion that *Dl* may be a complex locus. Three lethal ($Dl^{FE30}$, $Dl^{FE32}$ and $Dl^{B107}$) have been recovered with clear antimorphic effects, as shown by the fact that animals that carry any of these alleles in heterozygosity with a $Dl^+$ duplication still exhibit wing vein defects.

The *Dl* locus spans a stretch of approximately 25 kb of genomic DNA to which several *Dl* mutations have been mapped by Southern blot analysis [23]. The transcriptional organization of the *Dl* locus is not yet completely understood. The 25 kb of genomic DNA, where the mutants map, comprises at least three overlapping transcription units that exhibit very complex regulation (M. Haenlin, B. Kramatschek and J. A. Campos-Ortega, unpublished), reflecting the genetic complexity of the *Dl* locus [19]. One of them produces three major, largely overlapping poly $(A^+)$ RNAs of 5.4, 4.6 and 3.6 kb; since this transcription unit extends throughout the 25 kb of genomic DNA, and due to its pattern expression, we identify the three RNAs as the *Dl* products. The 5.4 kb RNA is zygotically expressed, whereas the other two are both maternal and zygotic. The second transcription unit encodes a maternally expressed 2.8 kb RNA; the third one encodes a 6 kb zygotic RNA. The relationship of these latter transcripts to *Dl* are unknown. *In situ* hybridization to embryonic tissue sections shows a distribution of the 5.4 kb RNA transcript that conforms with the expectation for a neurogenic gene [23]. Two main aspects of its very complex expression pattern should be emphasized: (i) the 5.4 kb RNA is expressed in territories with neurogenic capacities, like the neurogenic ectoderm or the anlagen of sensory organs; and (ii) after an initial phase, during which it is abundantly transcribed in all cells of such territories, the RNA becomes restricted to the cells that adopt the neural fate, e. g., the neuroblasts or the cells forming sensory organs, and persist in those cells for some time.

The sequence of the putative protein encoded by the 5.4 kb *Dl* transcript has been deduced from cDNA clones [23] and shows some similarity to the putative *N* protein [24–26]. The sequence of a 4.7 kb cDNA clone, which apparently encompasses all translated sequences of the major 5.4 kb *Dl* transcript, indicates a transmembrane protein with a number of features, among them a putative signal peptide, five potential glycosylation sites, and an extracellular domain comprising nine cysteine-rich tandem repeats with homology to various proteins of mammals, among them the epidermal growth factor (EGF) [24]. The primary structure of the putative *Dl* protein proposed by Vässin et al. [23] has been recently confirmed by Kopczynski et al. [27].

I would like to emphasize that important similarities between the proteins encoded by *D1* and by *N*, another neurogenic gene. The sequence of the 10.5 kb poly A$^+$ RNA transcribed from the *Notch* gene has been established [25, 26], and the conceptual translation of the sequence reveals the putative *N* product as a protein of 2703 amino acids. The primary structure of this protein is compatible with a transmembrane location. Its extracellular domain consists mainly of 36 EGF-like repeats. There are also three copies of another cysteine-rich repeated motif in the extracellular part, called the *Notch* repeats [25]. Antibodies directed against various parts of the *N* protein have permitted us to confirm its location at the membrane [28, 29]. Obviously, the structure of the *N* protein is compatible with its participation in cell communication processes, as suggested by embryological and genetic data. In view of their homology to the EGF, which is synthesized from a larger precursor molecule to be released [29], the repeats encoded by *N* and *D1* might also be conceivably cleaved from the cell membrane and diffuse through the intercellular space. However, data from genetic mosaics indicate that the products of both *N* and *D1*, as well as the products of the other NG genes, are unlikely to diffuse over long distance [16, 18]. It is more probable that these products mediate protein-protein interactions between neighboring, rather than distant, cells. The results of transplanting mutant cells [7] can be interpreted to suggest that both proteins act on the side of the signal source. Since *D1* shows the appropriate topological specificity in its expression, it seems a better candidate than the *N* protein, which is ubiquitously distributed, to mediate the protein-protein interactions between adjacent cells required for lineage segregation.

### 4.1.3.1.3 The Genetics and Molecular Genetics of the *Enhancer of Split* Locus

The *E(spl)* locus was discovered by means of the mutation *E(spl)*$^D$ [30]. The presence of this mutation in the genome enhances and renders dominant the expression of a mutation at the *N* locus called *split*. *E(spl)*$^D$ is homozygous viable and fertile, the homozygotes being virtually wild-type in phenotype; without the concomitant presence of *spl* in the genome, *E(spl)*$^D$/*E(spl)*$^D$ flies merely exhibit a slight compound eye roughening. However, 18 % of the embryos derived from these homozygotes develop neural hypoplasia defects of variable severity that affect structures of both the CNS and the PNS. Since all phenotypic traits of *E(spl)*$^D$ are enhanced, rather than suppressed, by additional copies of the *E(spl)*$^+$ allele in the genome, *E(spl)*$^D$ behaves as a gain of function mutation in which the gene product is modified, rather than absent [30].

An increasingly large body of experimental evidence [31–33] indicates that the *E(spl)* locus consists of a cluster of several related genetic functions rather than a single gene. First, evidence for the complex organization of the *E(spl)* locus derives from transmission genetics. Loss of function alleles of *E(spl)* can be easily recovered by reverting the dominant effect of $E(spl)^D$ upon *spl* [14]. As a rule, revertants associated with the loss of the *E(spl)* function are embryonic lethals and produce the neurogenic phenotype discussed above to a variable extent, depending on the allele; second, site revertants can also be recovered from the same experiments, and some of them correspond to $DI^-$ or $neu^-$ mutations, most probably reflecting the special functional interrelationships between NG genes [30, 32]. The frequency of recovery of loss of function *E(spl)* alleles as revertants of $E(spl)^D$ is relatively low, $1.5–2 \times 10^{-4}$, after irradiation of mature sperm with 5000 R, whereas *E(spl)* mutations can be recovered at a higher frequency ($2–3 \times 10^{-3}$) by other means. This low frequency, together with the fact that no revertant of $E(spl)^D$ has been obtained after EMS mutagenesis [32], suggests that the reversion of $E(spl)^D$ requires a complicated type of molecular lesion [see next section on the molecular characterization of $E(spl)^D$]. Indeed, most of the $E(spl)^D$ revertants are actually associated with chromosomal aberrations that permit it to locate the neurogenic defects of *E(spl)* loss of function mutations to the chromosomal bands 96F8–13.

There is an obvious correlation between the severity of the neural phenotype developed by homozygous embryos and the chromosomal aberration they carry, in that only large deletions lead to severe, fully penetrant neural hyperplasia, whereas homozygosity for inversions or translocations, or any of the other X-ray induced mutations, produce weak to intermediate phenotypes with incomplete penetrance only [31, 32, 34]. Expressivity and penetrance of the neurogenic phenotypes of all these mutations can be increased when the corresponding variants are heterozygous with large deletions of the 96F region. A number of lethal *E(spl)* alleles have been recovered after EMS mutagenesis [32]. None of them causes fully penetrant neural hyperplasia in homozygosity. Thus, with respect to neural hyperplasia, *E(spl)* alleles behave differently from alleles of the remaining NG loci, e. g., *N*, *mam*, or *D1*, in which several point mutants are known that lead to the amorphic phenotype of the corresponding gene. Consideration of these findings led us to postulate functional redundancy with respect to the role played by the *E(spl)* locus in neurogenesis. It appears that several functions have to be eliminated simultaneously in order to abolish the function of the locus completely [32].

*E(spl)* mutations exhibit a complex pattern of heteroallelic complementation, with respect to their viability in crosses with lethal alleles and to expression of visible phenotypic traits in crosses with $E(spl)^D$ and *gro*. All the data derived from the complementation analysis are indicative of considerable genetic complexity [32].

The results of the on-going molecular analysis confirm and extend the conclusion derived from genetic studies that the *E(spl)* locus is a gene complex encoding several related functions [31–33]. The composition of the *E(spl)* gene complex is not yet precisely defined; however, the current results show that the *E(spl)* gene corresponds to the transcription unit *m8* and that the *E(spl)* complex consists of at least the transcription units *m4*, *m5*, *m7*, *m8*, and *m9-m10* (see below). All of these functions have to be affected in order for the neurogenic process to be disturbed. The variant *Df(3)E(spl)*[R-A7.1], which lacks 34–36 kb of genomic DNA (ref. to Fig. 5), is defective for some *E(spl)* functions. Several mutations, including *E(spl)*[D], have been mapped to the same stretch of genomic DNA [31, 34]. However, it seems probable that homozygous *Df(3)E(spl)*[R-A7.1] embryos do not lack all the functions of the *E(spl)* locus because they do not exhibit the most severe form of the neurogenic phenotype. In fact, embryos homozygous for deletions bigger than *Df(3)E(spl)*[R-A7.1] develop a considerably more severe form of that phenotype. Thus, the *E(spl)* complex is likely to extend still further proximal and/or distal to the territory deleted in *Df(3)E(spl)*[R-A7.1].

The 34–36 kb genomic DNA deleted in *Df(3)E(spl)*[R-A7.1] to which *E(spl)* mutations were mapped contains at least ten different transcription units and encodes eleven transcripts (one of the units encodes two overlapping RNAs) that have been called *m1* to *m11*, in proximo-distal direction. Experimental evidence indicates that seven of the eleven RNAs may be related with *E(spl)* functions. Four of these RNAs are affected in mutations: *m3*, a 1.4 kb RNA, is missing in *T(3;4)E(spl)*[R-C1.4c] embryos and is larger, 1.6 kb, in *gro* embryos; *m9* and *m10* are more abundantly expressed in *E(spl)*[D] than in wild-type; and *m8* is shorter and more abundantly expressed in *E(spl)*[D] than in wild-type [31, 33]. Several molecular lesions have been identified in the genomic DNA of the *E(spl)*[D] allele, including a middle repetitive fragment inserted in the transcription unit *m9-m10*, and various deletions and insertions in the coding and 5′ regions of the *m8* transcription unit [31, 33]. P-element mediated transformation experiments, in which a mutant *m8* transcription unit, derived from the genome of *E(spl)*[D] animals, is injected into the germ line of wild-type animals, demonstrate that the transgenic flies exhibit the ability to enhance the phenotype of *spl* [33]. Thus, the molecular lesions associated with the *m8* transcription unit are responsible for the phenotype of the *E(spl)*[D] allele and, hence, the *m8* transcription unit corresponds to the *E(spl)* gene.

The evidence to support that transcription units *m4*, *m5*, *m7*, and *m8* form part of the *E(spl)* complex is still indirected and chiefly based on similarities in the pattern of expression and the sequence of the encoded products. On the one hand, all four RNAs show a very similar pattern of mRNA distribution during embryogenesis; this distribution conforms to the expectation for the epidermalizing function assumed to be exerted by *E(spl)*[+]. On the other hand, sequence analyses [33] have uncovered extensive sequence

157

José Antonio Campos-Ortega

homology in the putative proteins encoded by the transcription units *m5*, *m7*, and *m8* and thus substantiated the hypothesis that the various products of the complex perform similar functions. I would like to recall that results of genetic studies led to propose a high degree of redundancy in the various functions of the *E(spl)* complex [32], and the sequence data strongly support this hypothesis. Whereas the protein encoded by *m4* is rather acidic, those of *m5*, *m7*, and *m8* are fairly basic. In addition, the amino-terminal domain of the proteins encoded by the latter three genes exhibits sequence similarity to a region comprising a helix-loop-helix (HLH) motif [35] conserved in several different proteins, among them members of the *myc* family [36–39], several proteins involved in muscle development (*MyoD1*, *myogenin*, and others [40–43]), and two immunglobulin enhancer binding proteins [44]. Interestingly, the same HLH motif has been found in other proteins of *Drosophila*, like *twist* [45], *hairy* [46] and, more important for our present purposes, in the proteins encoded by the transcripts T2, T3, T5, and T1a of the AS-C [47–49] and in the *daughterless* [50] protein, all which are also required for neural development (see below).

Transformation experiments have also been used by Preiss et al. [34] to present evidence that the RNAs *m9-m10* are also part of the *E(spl)* complex. Hartley et al. [51] determined the sequence of the protein encoded by the overlapping transcripts *m9-m10* and showed that it is similar to the β subunit of transducin, a G protein of mammals. Along with the sequence similarity of the three small *E(spl)* proteins, *m5*, *m7*, and *m8*, to *myc* proteins, described above, this homology of the putative *m9-m10* protein is a most appealing one, for G proteins are intimately associated with receptors. The results of transplantation experiments using cells homozygous for NG mutations [7], in which only *E(spl)*⁻ cells showed an autonomous phenotypic expression, prompted us to propose the participation of *E(spl)* in functions related to the reception of the epidermalizing signal. These data suggest that the *E(spl)* locus may encode consecutive steps of this process, from the transduction of the signal at the cell membrane to the regulation of the genetic activity in the receiving cell.

4.1.3.1.4  The Genetics and Molecular Genetic of the Master Mind Locus

The *mam* locus (50C23-D1; [52,53]) has been characterized by 41 non-complementing recessive lethals, which were recovered from different mutagenesis programs ([13, 54]; H. Schrons, U. Wetter, D. Weigel, U. Dietrich and J. A. Campos-Ortega, unpublished). These alleles cause in neurogenic phenotype with variable expressivity, allowing the establishment of an allelic series [13, 14, 52]. Flies heterozygous with any of several amorphic *mam* mutations exhibit various defects of the wings, particularly notching at the posterior margin and delta-shaped widenings of the tips of the veins. These

158

phenotypic traits are remarkably similar to those of heterozygotes for *N* or *D1* amorphic mutations.

Very little is known about the molecular organization and expression of the *mam* locus. Genomic DNA from the *mam* locus has been cloned and partially characterized by Weigel et al. [52] and Yedvobnick et al. [53]. The limits of the locus are not yet well established. Several *mam* mutations have been mapped to a stretch of 45–60 kb of genomic DNA. This DNA contains a large number of copies of two different repetitive sequences. Sequence analysis [53] has shown that one of the repeats corresponds to *opa* and the other to the *N*- [52] or *RS*-repeat [53]. The 45 kb stretch encodes two major overlapping RNAs, of approximately 5.0 and 3.9 kb which show the expected temporal regulation, i. e., strong maternal expression and zygotic expression during 3 to 8 h of embryonic development. The abundant repetitive sequences render the further molecular analysis of the *mam* locus very difficult.

### 4.1.3.1.5  The Genetics and Molecular Genetics of the Neuralized Locus

The *neu* locus is one of the genetically less well characterized NG loci. *neu* has been located to 85C4-14 by means of various chromosomal aberrations (A. de la Concha and J. A. Campos-Ortega, unpublished) and further characterized by 15 non-complementing lethal alleles recovered from EMS and X-ray mutageneses. Ten of the alleles lead to strong neurogenic defects, whereas the other five only cause intermediate or weak phenotypes. The *neu*$^+$ locus is included in the *Dp(3;3)Antp*$^{+R8}$, and a few embryos derived from this strain exhibit neural hypoplasic defects, reminiscent of those found among the progeny of females triploid for *E(spl)*$^+$. It is remarkable that increasing the ploidy of *N*$^+$ or *D1*$^+$ does not cause such embryonic defects, suggesting particular relationships between *neu* and *E(spl)* (see below). *neu* also has a maternal component of expression [16].

The molecular analysis of the *neu* locus has been initiated with the molecular cloning of its DNA as result of a collaboration between my group and the groups of L. and Y. N. Jan at the Department of Biochemistry in the University of California San Francisco. The results of this work are in process of being published.

## 4.1.3.2  Functional Interactions Between the Neurogenic Loci

The identical phenotype caused by the loss of any of the neurogenic gene functions strongly suggests that the products of these gene functions participate in a single function pathway. A large number of observations have been made in this respect [55, 16, 21, 56], and they strongly indicate that all of the neurogenic loci tested, with the exception of *bib*, are in fact involved in a common pathway. Since mutations in most of the neurogenic loci exhibit dominant traits in their phenotypic expression, e. g. wing, bristle, or leg defects, some interactions were worked out using heterozygous flies; in other cases homozygous embryos were used to determine effects on neurogenesis.

### 4.1.3.2.1 Genetic Interactions in Heterozygotes

The observation that animals lacking one wild-type allele of the *Dl* and *E(spl)* loci do not survive [14] was soon extended by the finding that the viability of animals doubly heterozygous for amorphic *N* and *E(spl)* mutations is also highly impaired [21]. Some of the double heterozygous animals die as embryos developing weak neural hyperplasia. Such behavior is particularly striking, for double heterozygotes for *N* and *Dl* mutations with a normal complement of *E(spl)*$^+$ are fully viable, and implies that half normal levels of expression at neither *N* or *Dl* and *E(spl)*, simultaneously, are not sufficient for normal embryogenesis.

Results of various genetic combinations, including both deficiencies and duplications of *N*$^+$, *Dl*$^+$ and *E(spl)*$^+$, led us to postulate a network of reciprocal interactions with opposite character between these three loci, the meaning of which still remains unclear [21]. In addition, *Dl*$^-$ mutations can be recovered as second-site revertants of *E(spl)*$^D$ indicating important interrelationships between both loci. Functional interactions between the proteins encoded by *N* and *Dl* are strongly indicated by two *Dl* alleles that have been recently recovered as suppressors of *spl* [57]. Since the suppression of the *spl* phenotype by *Dl* is allelic specific, for it is not attained by *Dl* deletions nor by other amorphic alleles, the interaction between *N* and *Dl* is likely to be at the level of the proteins.

Interrelationships between *E(spl)* and *N* are indicated by the allele specific enhancement of the *spl* phenotype by the *E(spl)*$^D$ allele. As discussed above, it is possible that this interaction occurs between the proteins encoded by *m8* and *N*. Interrelationships between *E(spl)* and *neu* are indicated by the fact that *neu*$^-$ mutations can be recovered as second-site revertants of *E(spl)*$^D$ (see above). In addition, *neu*$^-$ mutations considerably reduce the enhancement of the *spl* phenotype caused by *E(spl)*$^D$, whereas increasing the ploidy of *neu*$^+$ leads to increased enhancement of *spl*. The described effects of *neu* upon *spl* are carried out chiefly through *E(spl)*, rather than directly on *N* itself, for *neu*$^-$ mutations exert very mild effects on *spl* when the *E(spl)*$^D$ allele is not present in the genome (A. de la Concha and J. A. Campos-Ortega, unpublished). However, this finding does not preclude the existence of other interactions directly between *neu* and *N*.

### 4.1.3.2.2  Genetic Interactions in Homozygotes

The use of duplications of the various neurogenic loci has permitted us to establish epistatic relationships between these loci. We observed that the homozygous phenotype of some of the neurogenic mutations can be modified when the genome carries an increased number of copies of the wild-type allele of another neurogenic locus [21, 56]. The analysis of embryos homozygous for two different neurogenic mutations, or being homozygous for a neurogenic mutation and carrying a duplication of the wild-type allele of another neurogenic locus, has led to postulate functional community for six neurogenic loci, whereas the seventh, *bib*, appears to be independent of the others. The results are consistent with these six loci being links of a chain or network of epistatic interactions the last link of which would be *E(spl)* locus.

## 4.1.3.3  Genetic Organization of the Subdivision 1B

The subdivision 1B of the X-chromosome contains several genes required for neurogenesis. Four genes which are related to bristle development and viability, i. e., *achaete (ac)*, *scute (sc)*, *lethal of scute (l'sc)* and *asense (ase)* constitute the *achaete-scute* gene complex (AS-C). The phenotypic analysis of the subdivision 1B mutants [58, 59] indicates that CNS defects can be related to the deletion of *l'sec*, on the one hand, and of *vnd*, on the other hand. With respect to its neurogenic functions, *l'sc* seems to interact with *ac, sc,* and *ase*; thus, the deletion of the four genes causes a phenotype of increased severity.

## 4.1.3.4  The Genetics and Molecular Genetics of the *Daughterless* Locus

The requirements for *da*$^+$ for normal neural development is a surprising finding [60], for the gene has been known for some time to be required for sex determinations and dosage compensation [61–63]. The locus is expressed both during oogenesis and embryonic development. There is evidence that the maternal expression is relevant at early blastoderm stages to obtain a correct dosage compensation and differentiation in female embryos, whereas the zygotic expression is essential for PNS development in both sexes (discussed in [60]). Loss of the *da*$^+$ function leads to embryonic lethality. The dead embryos show a predominantly neural phenotype: they lack all sensory neurones. In addition, the CNS is smaller than normal, the ventral cord being frequently fragmented in several pieces. The *da* phenotype is chiefly associated with cell death during stages 11–13 [57]; defects in the pattern of neuroblasts of *da* embryos can also be distinguished (M. Brand and J. A. Campos-Ortega, in preparation).

The *da* locus has been recently cloned in a collaboration between my laboratory and that of L. and Y. N. Jan at UCSF [50]. Five *da* mutations have been mapped within an interval of approximately 5 kb of genomic DNA, which encodes a single transcription unit with two overlapping RNAs of 3.2 and 3.7 kb. The conceptual translation of the corresponding cDNA sequences uncovers an interesting similarity, namely the conserved HLH region present in *myc*, in the proteins encoded by the AS-C transcripts T3, T4, T5, and T8 and in the proteins encoded by the *E(spl)* transcripts *m5, m7,* and *m8.*

## 4.1.3.5   Interactions Between Neurogenic and Genes of the AS-C and *da*

Double mutants [57] show that the severity of the phenotype of homozygous neurogenic mutants can be considerably reduced if a mutation of the AS-C or of *da* is present in homo- or hemizygosity in the same genome. This reduction of the phenotypic severity of the double mutant affects both the epidermis, which is larger, and the neural tissue, which is less hyperplasic. At least some of the interactions between neurogenic and AS-C genes are likely to involve an influence on the pattern of transcription of these genes. Changes of the pattern of transcription of the genes T3 and T5 (*l'sc* and *ac*) have been observed in embryos carrying any of several neurogenic mutations [57]. In these embryos, T3 and T5 are expressed in more cells than in the wild-type. However, the early pattern of expression, up to stage 9, of T3 and T5 in neurogenic mutants is indistinguishable from the wild-type; hence, transcriptional interactions seem to operate, or at least to become evident, at the time when the segregation of lineages is taking place. In contrast to this finding, no significant modification of the pattern of transcription of neurogenic genes is observed in $Df(1)sc^{19}$ embryos, which lack most of the AS-C genes [57]. Therefore, the polarity of functional relationships between the two gene groups is likely to be from the NG to the AS-C genes, and not *vice versa.* These results suggest that cellular interactions mediated by the neurogenic genes are responsible for the refinement of the territories of T3-T5 expression in the wild-type and that the neurogenic genes exert this function by suppressing the transcription of T3 and T5 in some of the neuroectodermal cells.

## 4.1.4   References

[1]   J. A. Campos-Ortega, V. Hartenstein: The embryonic development of Drosophila melanogaster. Springer-Verlag, Berlin – Heidelberg – New York – Tokyo 1985, 235 pages.

[2] V. Hartenstein, J. A. Campos-Ortega: Early neurogenesis in wildtype Drosophila melanogaster. Roux's Arch. Dev. Biol. 193 (1984) 308–325.

[3] G. M. Technau, J. A. Campos-Ortega: Fate mapping in wildtype Drosophila melanogaster. II. Injections of horseradish peroxidase in cells of the early gastrula stage. Roux's Arch. Dev. Biol. 194 (1985) 196–212.

[4] V. Hartenstein, J. A. Campos-Ortega: The pattern of proliferation of the neuroblasts in the wild-type embryo of Drosophila melanogaster. Roux's Arch. Dev. Biol. 196 (1987) 473–485.

[5] P. H. Taghert, C. Q. Doe, C. S. Goodman: Cell determination and regulation during development of neuroblasts and neurones in grasshopper embryos. Nature 307 (1984) 163–165.

[6] G. M. Technau, Campos-Ortega: Lineage analysis of transplanted individual cells in embryos of Drosophila melanogaster. II. Commitment and proliferative capabilities of neural and epidermal cell progenitors. Roux's Arch. Dev. Biol. 195 (1986) 445–454.

[7] G. M. Technau, J. A. Campos-Ortega: Cell autonomy of expression of neurogenic genes of Drosophila melanogaster. Proc. Natl. Acad. Sci. USA 84 (1987) 4500–4504.

[8] G. M. Technau, T. Becker, J. A. Campos-Ortega: Reversible commitment of neural and epidermal progenitor cells during embryogenesis of Drosophila melanogaster. Roux's Arch. Dev. Biol. 197 (1988) 413–418.

[9] J. A. Campos-Ortega: Cellular interactions during early neurogenesis of Drosophila melanogaster. Trends in Neurosc. 11 (1988) 400–405.

[10] J. A. Campos-Ortega: Genetics of early neurogenesis in Drosophila melanogaster. Trends in Neurosc. 8 (1985) 245–250.

[11] D. F. Poulson: Chromosomal deficiencies and embryonic development of Drosophila melanogaster. Proc. Natl. Acad. Sci. USA 23 (1937) 133–137.

[12] T. R. F. Wright: The Genetics of embryogenesis in Drosophila. Adv. Genetics. 15 (1970) 262–395.

[13] R. Lehmann, U. Dietrich, F. Jimenez, J. A. Campos-Ortega: Mutations of early neurogenesis in Drosophila. Roux's Arch. Dev. Biol. 190 (1981) 226–229.

[14] R. Lehmann, F. Jimenez, U. Dietrich, J. A. Campos-Ortega: On the phenotype and development of mutants of early neurogenesis in Drosophila melanogaster. Roux's Arch. Dev. Biol. 192 (1983) 62–74.

[15] F. Jimenez, J. A. Campos-Ortega: Maternal effects of zygotic mutants affecting early neurogenesis in Drosophila. Roux's Arch. Dev. Biol. 191 (1982) 191–201.

[16] U. Dietrich, J. A. Campos-Ortega: The expression of neurogenic loci in imaginal epidermal cells of Drosophila melanogaster. J. Neurogen. 1 (1984) 315–332.

[17] V. Hartenstein, J. A. Campos-Ortega: The peripheral nervous system of mutants of early neurogenesis in Drosophila melanogaster. Roux's Arch. Dev. Biol. 195 (1986) 210–221.

[18] P. E. Hoppe, R. J. Greenspan: Local function of the Notch gene for embryonic ectodermal choice in Drosophila. Cell 46 (1986) 773–783.

[19] H. Vässin, J. A. Campos-Ortega: Genetic analysis of Delta, a neurogenic gene of Drosophila melanogaster. Genetics 116 (1987) 433–445.

[20] D. L. Lindsley, L. Sandler, B. S. Baker, A. T. C. Carpenter, R. E. Denell, J. C. Hall, P. A. Jacobs, G. L. Miklos, B. K. Davis, R. C. Gethman, R. W. Hardy, A. Hessler, S. M. Miller, H. Nozawa, D. M. Parry, M. Gould-Somero: Segmental anaeuploidy and the genetic structure of the Drosophila genome. Gentics 71 (1972) 157–184.

[21] H. Vässin, J. Vielmetter, J. A. Campos-Ortega: Genetic interactions in early neurogenesis of Drosophila melanogaster. J. Neurogen. 2 (1985) 291–308.

[22] A. K. Alton, K. Fechtel, A. L. Terry, S. B. Meikle, M. A. T. Muskavitch: Cytogenic definition and morphogenetic analysis of Delta, a gene affecting neurogenesis in Drosophila melanogaster. Genetics 118 (1988) 235–245.

[23] H. Vässin, K. A. Bremer, E. Knust, J. A. Campos-Ortega: The neurogenic locus Delta of Drosophila melanogaster is expressed in neurogenic territories and encodes a putative transmembrane protein with EGF-like repeats. EMBO J. 6 (1987) 3431–3440.

[24] E. Knust, U. Dietrich, U. Tepaß, K. A. Bremer, D. Weigel, H. Vässin, J. A. Campos-Ortega: EGF-homologous sequences encoded in the genome of Drosophila melanogaster and their relation to neurogenic genes. EMBO J. 6 (1987a) 761–766.

[25] K. A. Wharton, K. M. Johansen, T. Xu, S. Artavanis-Tsakonas: Nucleotide sequence from the neurogenic locus Notch implies a gene product that shares homology with proteins containing EGF-like repeats. Cell 43 (1985a) 567–581.

[26] S. Kidd, M. R. Kelley, M. W. Young: Sequence of the Notch locus of Drosophila melanogaster: Relationship of the encoded protein to mammalian clotting and growth factors. Mol. Cell. Biol. 6 (1986) 3094–3108.

[27] C. C. Kopczynski, A. K. Alton, K. Fechtel, P. J. Kooh, M. A. T. Muskavitch: Delta, a Drosophila neurogenic gene, is transcriptionally complex and encodes a protein related to blood coagulation factors and epidermal growth factor of vertebrates. Genes & Dev. 2 (1988) 1723–1735.

[28] S. Kidd, M. K. Baylies, G. P. Gasic, M. W. Young: Structure and distribution of the Notch protein in developing Drosophila. Genes & Dev. 3 (1989) 1113–1129.

[29] A. Gray, T. J. Dull, A. Ullrich: Nucleotide sequence of epidermal growth factor cDNA predicts a 128,000-molecular weight protein precursor. Nature 303 (1983) 722–725.

[30] E. Knust, K. A. Bremer, H. Vässin, A. Ziemer, U. Tepaß, J. A. Campos-Ortega: The Enhancer of split locus and neurogenesis in Drosophila melanogaster. Dev. Biol. 122 (1987a) 262–273.

[31] E. Knust, K. Tietze, J. A. Campos-Ortega: Molecular analysis of the neurogenic locus Enhancer of split of Drosophila melanogaster. EMBO J. 6 (1987c) 4113–4123.

[32] A. Ziemer, K. Tietze, E. Knust, J. A. Campos-Ortega: Genetic analysis of Enhancer of split, a locus involved in neurogenesis in Drosophila melanogaster. Genetics 119 (1988) 63–74.

[33] C. Klämbt, E. Knust, K. Tietze, J. A. Campos-Ortega: Closely related transcripts encoded by the neurogenic gene complex Enhancer of split of Drosophila melanogaster. EMBO J. 8 (1989) 203–210.

[34] A. Preiss, D. A. Hartley, S. Artavanis-Tsakonas: The molecular genetics of Enhancer of split, a gene required for embryonic neural development in Drosophila. EMBO J. 7 (1988) 3917–3928.

[35] C. Murre, P. Schonleber McCaw, D. Baltimore: The amphipath helix-loop-helix: a new DNA-binding and dimerization motif in immunglobin enhancer binding, daughterless, MyoD and myc proteins. Cell 56 (1989) 777–783.

[36] R. Watt, L. W. Stanton, K. B. Marcu, R. C. Gallo, C. M. Croce, G. Rovera: Nucleotide sequence of cloned cDNA of human c-myc oncogene. Nature 303 (1983) 725–728.

[37] N. E. Kohl, E. Legouy, R. A. DePinho, P. D. Nisen, R. K. Smith, C. E. Gee, F. W. Alt: Human N-myc is closely related in organization and nucleotide sequence to c-myc. Nature 319 (1986) 73–77.

[38] J. Stone, T. DeLange, G. Ramsey, E. Jakobovits, J. M. Bishop, H. Varmus, W. Lee: Definition of regions in human c-myc that are involved in transformation and nuclear localization. Mol. Cell Biol. 7 (1987) 1697–1709.

[39] E. Legouy, R. dePinho, K. Zimmermann, R. Collum, G. Yancopoulos, L. Mitsock, R. Kriz, F. W. Alt: Structure and expression of the murine L-myc gene. EMBO J. 6 (1987) 3359–3366.

[40] S. J. Tapscott, R. L. Davis, M. J. Thayer, P.-F. Cheng, H. Weintraub, A. B. Lassar: MyoD1: a nuclear phosphoprotein requiring a myc homology region to convert fibroblasts to myoblasts. Science 242 (1988) 405–411.

[41] W. W. Wright, D. A. Sassoon, V. K. Lin: Myogenin, a factor regulating mygenesis has a domain homologous to MyoD. Cell 56 (1989) 607–617.

[42] D. G. Edmondson, E. N. Olson: A gene with homology to the myc similarity region of MyoD1 is expressed during myogenesis and is sufficient to activate the muscle differentiation program. Genes & Devel. 3 (1989) 628–640.

[43] T. Braun, G. Buschhausen-Denker, E. Bober, E. Tannich, H. H. Arnold: A novel human muscle factor related but distinct from MyoD1 induces myogenic conversion in 10T1/2 fibroblasts. EMBO J. 8 (1989) 701–709.

[44] C. Murre, P. Schonleber McCaw, H. Vässin, M. Caudy, L. Y. Jan, Y. N. Yan, C. V. Cabrera, J. N. Buskin, S. D. Hauschka, A. B. Lassar, H. Weintraub, D. Baltimore: Interactions between heterologous helix-loop-helix proteins generate complexes that bind specifically to a common DNA sequence. Cell 58 (1989) 537–544.

[45] B. Thisse, C. Stoetzel, C. Gorostiza-Thisse, F. Perrin-Schmitt: Sequence of the twist gene and nuclear localization of its protein in endomesodermal cells of early Drosophila embryos. EMBO J. 7 (1988) 2175–2183.

[46] C. A. Rushlow, A. Hogan, S. M. Pinchin, K. M. Howe, M. Lardelli, D. Ish-Horowicz: The Drosophila hairy protein acts in both segmentation and bristle patterning and shows homology to N-myc. EMBO J. 8 (1989) 3095–3103.

[47] R. Villares, C. V. Cabrera: The achaete-scute gene complex of Drosophila melanogaster: conserved domains in a subset of genes required for neurogenesis and their homology to myc. Cell 50 (1987) 415–424.

[48] M. C. Alonso, C. V. Cabrera: The achaete-scute gene complex of Drosophila melanogaster comprises four homologous genes. EMBO J. 7 (1988) 2585–2591.

[49] F. Gonzales, S. Romani, P. Cubas, J. Modolell, S. Campuzano: Molecular analysis of asense, a member of the achaete-scute complex of Drosophila melanogaster, and its novel role in optic lobe development. EMBO J. 8 (1989) 3553–3562.

[50] M. Caudy, H. Vässin, M. Brand, R. Tuma, L. Y. Jan, Y. N. Jan: daughterless, a gene essential for both neurogenesis and sex determination in Drosophila, has sequence similarities to myc and the achaete-scute complex. Cell 55 (1988) 1061–1067.

[51] D. A. Hartley, A. Preiss, S. Artavanis-Tsakonas: A deduced gene product from the Drosophila neurogenic locus Enhancer of split shows homology to mammalian G-protein β subunit. Cell 55 (1988) 785–795.

[52] D. Weigel, E. Knust, J. A. Campos-Ortega: Molecular Organization of master mind, a neurogenic gene of Drosophila melanogaster. Mol. Gen. Genet. 207 (1987) 374–384.

[53] B. Yedvobnick, D. Smoller, P. Young, D. Mills: Molecular Analysis of the neurogenic locus mastermind of Drosophila melanogaster. Genetics 118 (1988) 483–497.

[54] C. Nüsslein-Vollhard, E. Wieschaus, H. Kluding: Mutations affecting the pattern of the larval cuticle of Drosophila melanogaster. I. Zygotic loci on the second chromosome. Roux's Arch. Dev. Biol. 193 (1984) 267–282.

165

[55] J. A. Campos-Ortega, R. Lehmann, F. Jimenez and U. Dietrich: A genetic analysis of early neurogenesis in Drosophila. In: S. C. Sharma (ed.): Organizing principles of neural development. Plenum Press, New York-London 1984, pp. 129–144.

[56] A. de la Concha, U. Dietrich, D. Weigel, J. A. Campos-Ortega: Functional interactions of neurogenic genes of Drosophila melanogaster. Genetics 118 (1988) 499–508.

[57] M. Brand, J. A. Campos-Ortega: Two groups of interrelated genes regulate early neurogenesis in Drosophila melanogaster. Roux's Arch. Dev. Biol. 197 (1988) 457–470.

[58] F. Jiménez, J. A. Campos-Ortega: A region of the Drosophila genome necessary for CNS development. Nature 282 (1979) 310–312.

[59] F. Jiménez, J. A. Campos-Ortega: Genes in subdivision 1B of the Drosophila melanogaster X-chromosome and their influence on neural development. J. Neurogen. 4 (1987) 179–200.

[60] M. Caudy, E. H. Grell, C. Dambly-Chaudiére, A. Ghysen, L. Y. Jan, Y. N. Yan: The maternal sex determination gene daughterless has zygotic activity necessary for the formation of peripheral neurons in Drosophila. Genes Dev. 2 (1988) 843–852.

[61] T. W. Cline: A sex specific temperature sensitive maternal effect of the daughterless mutation of Drosophila melanogaster. Genetics 84 (1976) 723–742.

[62] T. W. Cline: Maternal and zygotic sex-specific gene interactions in Drosophila. Genetics 96 (1980) 903–926.

[63] J. C. Luchesy, T. Skripsy: The link between dosage compensation and sex differentiation in Drosophila melanogaster. Chromosoma. 82 (1981) 217–227.

## 4.1.5   Publications from 1985 to 1988

J. Beer, G. M. Technau, J. A. Campos-Ortega: Lineage analysis of transplanted individual cells in embryos of Drosophila melanogaster. IV. Commitment and proliferative capabilities of mesodermal cells. Roux's Arch. Dev. Biol. 19 (1987) 222–230.

M. Brand, J. A. Campos-Ortega: Two groups of interrelated genes regulate early neurogenesis in Drosophila melanogaster. Roux's Arch. Dev. Biol. 197 (1988) 457–470.

J. A. Campos-Ortega: Genetics of early neurogenesis in Drosophila melanogaster. Trends in Neurosc. 8 (1985) 245–250.

J. A. Campos-Ortega: Cellular interactions in early neurogenesis of Drosophila melanogaster. Trends in Neurosc. 11 (1988) 400–405.

J. A. Campos-Ortega, K. A. Bremer, A. de la Concha, U. Dietrich, E. Knust, G. M. Technau, K. Tietze, H. Vässin, A. Ziemer: Genetische Analyse der frühen Neurogenese bei Drosophila. Verh. Dtsch. Zool. Ges. 80 (1988) 9–22.

J. A. Campos-Ortega, V. Hartenstein: The embryonic development of Drosophila melanogaster. viii+227 pages, Springer-Verlag, Berlin-Heidelberg-New York-Tokyo 1985, 235 pages.

M. Caudy, H. Vässin, M. Brand, R. Tuma, L. Y. Jan, Y. N. Jan: Daughterless, a gene essential for both neurogenesis and sex determination in Drosophila, has sequence similarities to myc and the achaete-scute complex. Cell 55 (1988) 1061–1067.

A. de la Concha, U. Dietrich, D. Weigel, J. A. Campos-Ortega: Functional interactions of neurogenic genes of Drosophila melanogaster. Genetics. 118 (1988) 499–508.

V. Hartenstein: The influence of segmental compartmentalisation on the development of the larval peripheral nervous system in Drosophila melanogaster. Roux's Arch. Dev. Biol. 196 (1986) 308–325.

V. Hartenstein, J. A. Campos-Ortega: Fate mapping in wildtype Drosophila melanogaster I. The pattern of embryonic cell divisions. Roux's Arch. Dev. Biol. 194 (1985) 181–195.

V. Hartenstein, J. A. Campos-Ortega: The peripheral nervous system of mutants of early neurogenesis in Drosophila melanogaster. Roux's Arch. Dev. Biol. 195 (1986) 210–221.

V. Hartenstein, E. Rudloff, J. A. Campos-Ortega: The pattern of proliferation of the neuroblasts in the wild-type embryo of Drosophila melanogaster. Roux's Arch. Dev. Biol. 196 (1987) 473–485.

V. Hartenstein, G. M. Technau, J. A. Campos-Ortega: Fate mapping in wildtype Drosophila melanogaster. III. A fate map of the blastoderm. Roux's Arch. Dev. Biol. 194 (1985) 213–216.

F. Jimenez, J. A. Campos-Ortega: Genes in subdivision 1B of the Drosophila melanogaster X-chromosome and their influence on neural development. J. Neurogen. 4 (1987) 179–200.

C. Klämbt, E. Knust, K. Tietze, J. A. Campos-Ortega: Closely related transcripts encoded by the neurogenic gene complex Enhancer of split of Drosophila melanogaster. EMBO J. 8 (1989) 203–210.

E. Knust, K. A. Bremer, H. Vässin, A. Ziemer, U. Tepaß, J. A. Campos-Ortega: The Enhancer of split locus and neurogenesis in Drosophila melanogaster. Dev. Biol. 122 (1987) 262–273.

E. Knust, U. Dietrich, U. Tepaß, K. A. Bremer, D. Weigel, H. Vässin, J. A. Campos-Ortega: EGF-homologous sequences encoded in the genome of Drosophila melanogaster and their relation to neurogenic genes. EMBO J. 6 (1987) 761–766.

E. Knust, K. Tietze, J. A. Campos-Ortega: Molecular analysis of the neurogenic locus Enhancer of split of Drosophila melanogaster. EMBO J. 6 (1987) 4113–4123.

G. M. Technau: Lineage analysis of transplanted individual cells in embryos of Drosophila melanogaster. I. The method. Roux's Arch. Dev. Biol. 194 (1986) 196–212.

G. M. Technau: A single cell approach to problems of cell lineage and commitment during embryogenesis of Drosophila melanogaster Development 100 (1987) 1–12.

G. M. Technau, T. Becker, J. A. Campos-Ortega: Reversible commitment of neural and epidermal progenitor cells during embryogenesis of Drosophila melanogaster. Roux's Arch. Dev. Biol. 197 (1988) 413–418.

G. M. Technau, J. A. Campos-Ortega: Fate mapping in wildtype Drosophila melanogaster. II. Injections of horseradish peroxidase in cells of the early gastrula stage. Roux's Arch. Dev. Biol. 194 (1985) 196–212.

G. M. Technau, J. A. Campos-Ortega: Lineage analysis of transplanted individual cells in embryos of Drosophila melanogaster. II. Commitment and proliferative capabilities of neural and epidermal cell progenitors. Roux's Arch. Dev. Biol. 195 (1986) 445–454.

G. M. Technau, J. A. Campos-Ortega: Lineage analysis of transplanted individual cells in embryos of Drosophila melanogaster. III. Commitment and proliferative capabilities of pole cells and midgut progenitors. Roux's Arch. Dev. Biol. 195 (1986) 489–498.

G. M. Technau, J. A. Campos-Ortega: Cell autonomy of expression of neurogenic genes of Drosophila melanogaster. Proc. Natl. Acad. Sci. USA 84 (1987) 4500–4504.

H. Vässin, K. A. Bremer, E. Knust, J. A. Campos-Ortega: The neurogenic locus Delta of Drosophila melanogaster is expressed in neurogenic territories and encodes a putative transmembrane protein with EGF-like repeats. EMBO J. 6 (1987) 3431–3440.

H. Vässin, J. A. Campos-Ortega: Genetic analysis of Delta, a neurogenic gene of Drosophila melanogaster. Genetics 116 (1987) 433–445.

H. Vässin, J. Vielmetter, J. A. Campos-Ortega: Genetic interactions in early neurogenesis of Drosophila melanogaster. J. Neurogen. 2 (1985) 291–308.

D. Weigel, E. Knust, J. A. Campos-Ortega: Molecular organization of master mind, a neurogenic gene of Drosophila melanogaster. Mol. Gen. Genet. 207 (1987) 374–384.

A. Ziemer, K. Tietze, E. Knust, J. A. Campos-Ortega: Genetic analysis of Enhancer of split, a locus involved in neurogenesis in Drosophila melanogaster. Genetics 119 (1988) 63–74.

# Survey of Doctoral Students and Postdoctoral Scientists

## Doctoral Students

Brand, Michael[*] (1986–1989): Genetik eines zellulären Differenzierungsschrittes: Wechselwirkungen zwischen neurogenen und proneuralen Genen während der frühen Neurogenese bei Drosophila melanogaster.

Bremer, Kirsten Anne[*] (1984–1989): Untersuchungen zur Struktur und Transkription des Delta-Locus bei Drosophila melanogaster.

Hartenstein, Volker[*] (1983–1986): Über die Struktur und Entwicklung des larvalen peripheren Nervensystems von Drosophila melanogaster.

Vässin, Harald[*] (1983–1987): Genetische Charakterisierung und molekulare Klonierung des Delta-locus von Drosophila melanogaster.

Wille, Christiane[*] (1984–1986): Immunologische, biochemische und molekulargenetische Charakterisierung eines neuralen Antigens aus Drosophila melanogaster.

## Postdoctoral Scientists

Amador de la Concha, Dr. (1985–1987), Universidad de Murcia.

Marc Haenlin, Dr. (1987–1989), LGME-CRNS, Strasbourg, France.

Fernando Jimenez, Dr. (1983–1987), Universidad Autonoma, Madrid, Spain.

Christian Klämbt, Dr. (1987–1988), Universität Freiburg, FRG.

Elisabeth Knust, Dr. (1983–1988), Universität Erlangen, FRG.

Gerhard M. Technau, Dr. (1983–1987), Universität Würzburg, FRG.

## Habilitations

Dr. Gerhard M. Technau (1987), Professor, Institut für Genetik, Johannes-Gutenberg-Universität Mainz, FRG.

Dr. Elisabeth Knust (1988), Privatdozentin, Institut für Entwicklungsphysiologie, Universität zu Köln, FRG.

(* Diploma in C.-O.'s laboratory)

# 4.2 The Function of Specific Nuclear Proteins in the Regulation of Gene Expression in *Drosophila*

Harald Saumweber*

## 4.2.1 Introduction

Polytene chromosomes provide the unique possibility to study the structural arrangement of the genome during interphase at a light microscopic level. This was used in the past in order to localize mutations by visual inspection, to map the location of cloned genes by *in situ* hybridization, or to isolate genes by direct manual dissection of chromosomes. In combination with sensitive immunological methods, polytene chromosomes also proved to be useful for the study of the distribution of several chromosomal proteins. Our aim was to identify *Drosophila melanogaster* nuclear proteins, which are involved in the regulation of gene expression. At the very end, we wanted to make use of polytene chromosomes as a model to study the distribution of these proteins on gene families and to follow their mutual interaction by direct light microscopic inspection at different developmental stages and in different tissues.

As a general approach we used the clonal nature of the immune response and hybridoma technology to "immunologically dissect" the complex mixture of proteins which reside in the nucleus. We established libraries of monoclonal antibodies against fractionated nuclear proteins of *Drosophila* tissue culture cells and chromatin fractions of *Drosophila* embryonic nuclei [1, 2]. The antibodies were screened by examining the distribution of their antigens with indirect immunofluorescence, on both polytene chromosomes and in histological preparations of the whole organism at various developmental stages [3, 4]. Candidate regulatory proteins thus identified were further characterized and quantified biochemically. Under carefully controlled conditions, the sensitivity of our assays was sufficient to detect antigens with frequencies of $10^3$ copies per diploid cell or some hundred molecules per site on an average polytene chromosome.

---

* Institut für Entwicklungsbiologie der Universität zu Köln

Of the 80 distinct antigens detected by this approach, three have been selected for further studies. One of the proteins was found on only a few sites on polytene chromosomes; however, it was very prominent on the telomeres of chromosome arms 2L and 3L in certain *Drosophila* laboratory stocks. Two other chromatin associated antigens were detected in a number of transcriptionally active sites (puffs) on polytene chromosomes. The binding of these proteins on chromosomes was investigated in somewhat more detail. In order to learn more about these proteins, we cloned and characterized their corresponding genes. For one of the puff-specific genes we also obtained mutations, which should further help to understand its *in vivo* function(s).

## 4.2.2    Puff Specific Proteins

### 4.2.2.1    Mapping the Binding Sites of Bj6 and Bx42 Proteins *in situ*

(by H. Saumweber)

Both Bj6 protein (82 kD) and Bx42 protein (66 kD) cosedimented with nucleosomes when RNAse digested chromatin was separated on sucrose gradients. Both chromatin bound proteins were detected in a set of puffs and prominent interbands on polytene chromosomes. The distribution of both proteins changed reproducibly, according to the developmental stage of the larva. The Bx42 protein was detected at 60–70 % of the loci (120 of 5000 possible loci), where Bj6 protein could be detected (200 loci) [5]. Both proteins may have different functions in the regulation of gene expression worthwhile for further studies.

The salivary gland secretion gene Sgs-4 at 3C11–12 was used as a model site to study these proteins in more detail. This X-chromosomal gene is puffed and transcribed up to the stage of massive ecdyson release, when both the puff regresses and the transcription ceases towards the end of third larval instar. The protein product is used later to fix the pupal case on a solid substrate. As could be shown by immunofluorescence, both proteins were present at 3C11–12 as long as the gene was active. Immediately after puff regression, the Bx42 protein was not detectable anymore, and the Bj6 protein was detected only in a very reduced amount. I used available deletions and P-element transformed strains to map the binding region of both proteins at the Sgs-4 locus in indirect immunofluorescence [6]. By P-element transformation, 2.6 kb of DNA from upstream of the start of Sgs-4 transcription was inserted at 3A which is located next to the parental locus at 3C11–12 on the X-chromosome. It could be shown that the transposed

DNA directs the binding of both Bj6- and Bx42 protein to this new location. A *Drosophila* strain carrying a small 52 bp deletion within these 2.6 kbp still showed the normal binding of Bj6 protein; however, the binding of Bx42 protein was largely reduced or not detectable at all. Thus the binding of the two puff-specific proteins at the Sgs-4 gene is different and the binding of the Bx42 protein is specifically dependent on this 52 bp sequence. This was corroborated by the observation that an adjacent deletion of 100 bp did not significantly change the binding of Bx42 protein. The 52 bp deletion overlaps a site which is specifically hypersensitive to DNAseI digestion of chromatin in salivary glands of third instar larvae. Such hypersensitive sites have been reported to reflect the binding of regulatory proteins in several other systems.

## 4.2.2.2 Structure and Expression of the Bj6 Gene
(by Hans von Besser, Petra Schnabel and Elke Fritz)

The antibodies can be used to isolate cDNA clones from cDNA expression libraries according to established methods. For this purpose, a polyclonal antiserum against immunopurified Bj6 protein was raised in mice. A lambda gt-11 cDNA library from 0–16 hr *Drosophila* embryos made by B. Hovemann (ZMB Heidelberg) contained cDNA clones coding for the Bj6 epitope. To show that we had cloned the Bj6-gene, the isolated cDNA was expressed in frame with bacterial sequences and the fusion protein obtained was immunized into mice. When these mouse antisera were used in indirect immunofluorescence and on Western blots they detected the Bj6 protein of *Drosophila* tissues in the same way as the original Bj6 monoclonal antibody.

On Northern blots the Bj6 gene showed two transcripts of 2.8 and 3.9 kb in length, which were both expressed throughout development. However, there was a peak of expression from blastoderm to early germ band elongation stages. The protein probably has a long half-live, as could be concluded from a comparison of protein and RNA expression profiles during early development. *In situ* hybridization with labeled anti-sense probes showed that the Bj6 gene was ubiquitously expressed in all tissues of the developing embryo.

Nearly full length cDNA clones were isolated and sequenced. These showed a single open reading frame of 2.1 kb coding for a basic, protein of 700 AA. In the aminoterminus there were stretches of glycin (sometimes alternating with asparagine and glutamine) similar to the so called *pen*-repeats of single strand nucleic acid binding proteins. Directly adjacent were two copies of a sequence motif with significant (30 %) homology to the so called RNP consensus sequence found in many proteins with a possible function in RNA packaging and splicing. The putative start of transcription was identified, as well as 2 polyadenylation signals, which were consistent with the two detected transcripts.

It was shown that Bj6 protein was bound to chromatin even following extensive RNAse digestion [2, 5]. Yet the sequence similarity to RNA-binding proteins prompted us to reinvestigate the RNA-binding properties of Bj6 protein. These studies are still in progress, but there is evidence to show that Bj6 protein has some preference to binding to poly-dA-dT double stranded DNA over its binding to RNA. Thus, the RNP-consensus motif could either reflect additional functionally important RNA-binding sites on a DNA binding protein, or these motifs are common to an yet unknown family of nucleic acid binding proteins.

Genomic Southern blots show that Bj6 is a single copy gene, and by *in situ* hybridization, its chromosomal location was determined in 14C1 on the X-chromosome. Homologous sequences occur in all *Drosophila* species studied so far, and we also obtained strong hybridization in *Hydra viridis* and *Caenorhabditis elegans*. However, with the vertebrate genomes studied, so far no hybridization has been obtained under conditions of high stringency. Further studies under low stringency conditions will be used in a more systematic approach to look at the conservation of this gene in the animal kingdom. This will give us a further hint of the functional importance of this gene. The cross-reaction of the Bj6 monoclonal antibody is restricted to *Drosophila*. However, this would not be contradictory to a more general conservation of the protein, since monoclonal antibodies in general are limited to a single restricted epitope of the whole protein molecule.

44 kb of genomic DNA including the Bj6 gene were isolated and mapped by restriction digestion. A comparison of genomic and cDNA sequences showed the presence of four introns. The knowledge of the genomic gene structure will allow us to study the sequences needed for the expression of the Bj6 gene and to analyze alterations induced by mutations (see 4.2.2.3).

## 4.2.2.3  Isolation of Bj6 Mutations
### (by Ralf Stanewsky)

The Bj6 gene could be localized precisely between 14B17 and 14C4 by the proximal breakpoints of two existing deletions. Df(1)l32 removed all Bj6 related sequences, whereas Df(1)r$^{D1}$ still left them unchanged. Screening for lethal effects, we nearly saturated the X-chromosomal region, defined by the two breakpoints, with X-ray and EMS mutagenesis. Three lethal complementation groups were defined by genetic analysis within this interval. One allele in one of the three complementation groups was shown to remove 2.7 kb DNA upstream of the start of Bj6-transcription. Since the 3' breakpoint of this deletion fell within 300 bp of the start of transcription, this deletion could remove sequences important for Bj6 expression and hence be responsible for the lethal (Bj6) phenotype. Since the Bj6 protein was shown to interact with many chromosomal sites, an early embryonic lethal phenotype had been expected, if this interaction would be of functional importance for the expression of the genes located in these sites. Hemi- or homozygous mutant animals die as first instar larvae and are probably unable to undergo the first larval molt. The relatively long survival is probably caused by the presence of both maternal Bj6 RNA and maternal Bj6 protein. The involvement of the Bj6 protein is further supported by the long half live of the protein, which still could be detected in hemi- and homozygous mutant embryos. To analyze the phenotype in the complete absence of the Bj6 protein, we currently generate germ line clones to remove the maternal effect. This will give us valuable information about what gene(s) are most sensitive in early development to the dysfunction of the Bj6 protein. In any case, the experiments show that the Bj6 protein has a vital function which cannot be replaced by other *Drosophila* proteins. We further demonstrated that by using these methods, it is possible to follow the classical route of genetics in a reversed order, starting with a specific antibody, against a nuclear protein up to the isolation of a mutation.

## 4.2.3  Puff Specific Proteins: Structure and Expression of the Bx42 Gene

(by Simone Mann)

In a similar screen with the Bx42 monoclonal antibody (compare 4.2.2.2), several cDNA clones were isolated, which coded for the Bx42 epitope. Polyclonal and monoclonal antibodies against bacterial fusion proteins detected the Bx42 antigen from *Drosophila* nuclei on Western blots; yet so far, these appear to detect the denatured protein only. The gene is located in 8C7 on the X-chromosome next to the proximal breakpoint of the deficiency Df(1) KA14, and it appears to be present in the genome as a single copy. We have already initiated a chromosomal walk to isolate the genomic region.

The gene is expressed continuously with a peak from late blastoderm to early germ band elongation, similar to the Bj6 expression. Again, we find two transcripts, which according to sequencing data may be explained by differential polyadenylation. Sequencing is not finished yet; however, it is apparently a very basic protein. According to this property, some homology to histone proteins has been suggested, but this has to await further analysis.

## 4.2.4   The Telomere Associated Protein At5
(by Barbara Guhr)

The At5 antibody detected two polypeptides at 55 and 57 kD molecular weight on immunoblots of the *Drosophila* tissue culture cell line Kc. Further (cytoplasmic) polypeptides with different profiles of expression were detected in *Drosophila* embryos. On polytene chromosomes from salivary glands, the antigens were located in about 20 chromosomal sites, which were very prominent in the telomere of the chromosome arm 2L, and in some laboratory strains, also in the telomere of chromosome arm 3L. To our knowledge, no telomeric protein of this specificity has been described so far, and not much is known about the protein composition and the structure of chromosome ends. The distribution of the antigens was reminiscent of the distribution of a repeated DNA sequence found 3 kbp DNA upstream of a *Drosophila* gene controlling growth of neural tissues l(2)gl (lethal (2) giant larvae; O. Schmidt pers. communication). This gene was mapped next to the telomere of chromosome 2L; however, DNA subclones of the more upstream region containing the repeated sequence (about 10 kbp in size) also showed hybridization at the telomere of chromosome 3L.

The signal at 3L showed a similar variability in different laboratory stocks, as had been observed by immunostaining with antibodies directed against the At5 protein. Strains, mutant in the l(2)gl gene, showed a dual behavior in immunostaining: deletion mutants removing the repeated sequence and more distal sequences along with the gene function (e. g l(2)gl 338) did not show any signal in the homolog carrying the deletion, when tested in hetero-zygous condition. However, rearrangements in the l(2)gl gene, which did not remove the repeat, showed detectable signals in immunostaining on both homologues in a heterozygote (e. g. l(2)gl 275). Although we do not know the distal breakpoint in l(2)gl 338, the evidence available suggests a binding of the At5 protein in the region of the repeated DNA sequences. We will test whether this repeat contains the binding site for the telomere specific anti-gen by more direct DNA binding assays.

Attempts to clone the gene coding for the At5 protein from cDNA expres-sion libraries resulted in the isolation of cDNA clones coding for the 57/55 kD proteins. This was shown on Western blots, using polyclonal and mono-clonal antibodies against a bacterial fusion protein, which was coded in part by this cDNA.

The research described in this report was supported from 1/86– 12/88 by the Sonderforschungsbereich of the Deutsche Forschungsgemeinschaft, SFB-74, J2, H. Saumweber.

174

# 4.2.5    References

[1]  H. Saumweber, P. Symmons, R. Kabisch, H. Will, F. Bonhoeffer: Monoclonal anti-
     bodies against chromosomal proteins of *Drosophila melanogaster*. Chromosoma 80
     (1980) 253–275.
[2]  M. Frasch: Charakterisierung chromatinassoziierter Kernproteine von *Drosophila
     melanogaster* mit Hilfe monoklonaler Antikörper. Dissertation Eberhard-Karls-
     Universität Tübingen 1985.
[3]  R. Dequin, H. Saumweber, J. W. Sedat: Proteins shifting from the cytoplasm into
     the nuclei during early embryogenesis of *Drosophila melanogaster*. Dev. Biol. 104
     (1984) 37–48.
[4]  M. Frasch, D. M. Glover, H. Saumweber: Nuclear antigens follow different path-
     ways into daughter during mitosis in early *Drosophila* embryos. J. Cell. Sci. 82
     (1986) 155–172.
[5]  M. Frasch, H. Saumweber: Two proteins from *Drosophila* nuclei are bound to chro-
     matin and are detected in a series of puffs on polytene chromosomes. Chromosoma
     97 (1989) 272–281.
[6]  H. Saumweber, G. Korge, M. Frasch: Two puff-specific proteins bind within the
     2.5 α kb upstream region of the *Drosophila melanogaster Sgs-4* gene. Chromosoma
     99 (1990) 52–60.

# 4.2.6    Publications from 1986 to 1988

M. Hochstrasser, D. Mathog, Y. Gruenbaum. H. Saumweber, J. W. Sedat: Spatial orga-
    nization of chromosomes in the salivary gland nuclei of Drosophila melanogaster.
    J. Cell. Biol. 102 (1986) 112–123.
H. Saumweber: Arrangement of chromosomes in interphase cell nuclei. In: W. Hennig
    (ed.): Results and Problems in Cell Differentiation 14, Structure and Function of
    Eucaryotic Chromosomes. Springer-Verlag, Berlin-Heidelberg 1987, pp. 223–234.
W. G. F. Whitfield, S. E. Millar, H. Saumweber, M. Frasch, D. Glover: Cloning of a gene
    encoding an antigen associated with the centrosome in *Drosophila*. J. Cell Sci. 89
    (1988) 467–480.
M. Frasch, M. Paddy, H. Saumweber: Developmental and mitotic behaviour of two
    novel groups of nuclear envelope antigens of *Drosophila melanogaster*. J. Cell Sci.
    90 (1988) 247–263.
Y. Gruenbaum. Y. Landesman, B. Drees, J. W. Bare, H. Saumweber, M. R. Paddy, J. W.
    Sedat, D. E. Smith, B. M. Benton, P. A. Fisher: *Drosophila* nuclear lamin precursor
    $DM_O$ is translated from either of two developmentally regulated mRNA species
    apparently encoded by a single gene. J. Cell Biol. 106 (1988) 585–596.
M. Hochstrasser, D. Mathog, Y. Gruenbaum, H. Saumweber, J. W. Sedat: Three-dimen-
    sional organization of interphase chromosomes in polytene nuclei of *Drosophila
    melanogaster*. In: G. Kahl (ed.): Architecture of eukaryotic genes. VCH, Weinheim
    1988, pp. 473–487.

M. Frasch, H. Saumweber: Two proteins from *Drosophila* nuclei are bound to chromatin and are detected in a series of puffs on polytene chromosomes. Chromosoma (Berl.) 97 (1989) 272–281.

H. Saumweber, G. Korge, M. Frasch: Two puff-specific proteins bind within the 2.5 k. b upstream region of the *Drosophila melanogaster Sgs-4* gene. Chromosoma 99 (1990) 52–60.

## Doctoral Students

Barbara Guhr (1985– 1988): Characterization of a chromosomal protein with specificity for telomeres of Drosophila melanogaster. Bayer AG, Leverkusen, FRG.

# 4.3 Differentiation, Pattern Formation, and Sexual Induction in *Volvox carteri*

Lothar Jaenicke and Rainer Gilles*

## 4.3.1 Introduction

Our topic is the biochemical mechanism of cell differentiation, in particular that of germ line cells. The model system is the multicellular green flagellate *Volvox carteri*, which forms spheroids consisting of only two types of cells: germ line cells, called gonidia, and biflagellated somatic cells [1]. By the molecular signal of a pheromone, the differentiation of the gonidia, which serve as germ cells in vegetative multiplication, is shifted to the formation of male and female gametes for sexual reproduction. This induction of sexual differentiation is a completely autonomous, highly specific process enabling the organism to survive under adverse conditions and providing for genetic recombination. It offers the cell biologist the opportunity to study the following in a paradigmatic manner: (i) Communication between organisms – the sexual inducer is a species-specific signal released by only one sex, the male. (ii) Signal transduction between two different cell types – the sexual inducer acts on both, gonidia and somatic cells; the latter emit induction-specific chemical signals, which may be answered by the germ cells. (iii) Transmembrane signalling – the sexual inducer, bound to receptors, elicits intracellular second messages. Since *Volvox* is a protist at the threshold of evolution to multicellularity, these studies simultaneously could shed light on the origin and evolution of hormone-like systems and cell signalling in multicellular organisms.

## 4.3.2 Morphological Basis

A *Volvox* spheroid consists of two differentiated cell types: 2000 biflagellated terminally differentiated somatic cells form a monolayered globular hull filled with a gel-like matrix in which are embedded up to 16 reproductive cells (gonidia) in a highly ordered pattern (Fig. 4.3.1).

---

* Institut für Biochemie der Universität zu Köln

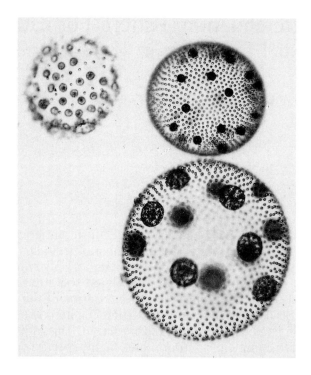

**Fig. 4.3.1:** *Volvox carteri* f. *nagariensis.* Vegetative spheroid (left), male and female spheroids (upper and lower right, respectively). Note the ordered pattern of the gonidia, sperm packets, and eggs.

This pattern is caused by the sequence of cleavage planes during ontogenesis. In the vegetative life cycle, the gonidia cleave five times to form an embryo with 32 blastomeres of equal size. In the next round, the 16 cells at the anterior pole cleave unequally; the large cells are the gonidial initials, while the small cells, as well as the rest of the cells, cleaving equally, divide five more times to produce the somatic cells. After an inverting process, which turns the embryo inside out, the young daughter spheroids grow, mature, and eventually hatch from the mother spheroid by proteolytic disintegration of the parental hull [2]. *Volvox carteri* does not always contain 16 gonidia but reduces the number according to the growth conditions. We found that these pattern variations are the result of premature by differentiating cleavages in organisms, already at the cell 16 stage [3].

Under the influence of the sexual inducer, the sexual pathway of development is initiated in both sexes. That is, the unequal cleavage is delayed in the female strain by one cell division, resulting in approximately 32 eggs; in the male strain, all cells cleave unequally only at the final division, which usually is the 256 cell stage. The unequal cell division preceding, and presumably determining, the soma/germ line differentiation must be controlled by temporal and spatial signals in the embryo. *Volvox* seems to have a counting mechanism for its cell cycles, and the time of the unequal, fate-determining division is influenced by the pheromonal signal and by other environmental conditions. Furthermore, a polarity exists defining the longitudinal axis of the embryo and its phialopore at the anterior pole. During vegetative and female development, only the cells at this end cleave unequally, while during male development, the polarity is lost and all cells undergo asymmetric cell division. We have presented a working hypothesis in which two types of molecules on the gonidial membrane, one gradedly distributed differentiation signal (D-factor) and an evenly distributed counting signal (C-factor), determine the spacing and timing of the unequal cell divisions [4]. Variation of the concentration of these signals at the beginning of embryogenesis could explain the patterns of reproductive cells in vegetative and sexual developmental pathways.

The sexual inducer has at least two effects during embryogenesis. First, it changes the time and pattern of the unequal cleavages, which differentiate germ and somatic cell lines (C-factor). This alone, however, is not sufficient to produce gametes. Several sterile mutants are known in which the unequal cleavage pattern is altered but which never produce gametes instead of gonidia. The inducer determines the differentiation of the germ line initials to eggs in the female and to androgonidia in male strains (D-factor). After embryogenesis the eggs rest for three days waiting for fertilization. If this does not happen, they begin cleaving parthogenetically. In male strains, the inducer switches on a completely new developmental program leading to sexual individuals. The androgonidia formed by the unequal cleavage will soon undergo spermatogenesis, and each will form a sperm packet with (usually) 64 sperm. As one of the late events in male development, the gene of the sexual inducer is activated, and the sexual inducer is secreted into the culture medium by the mature sperm [5, 6].

## 4.3.3    Sex$^c$-Switch

One surprising observation on growing *Volvox carteri* is the spontaneous induction of male cultures, which regularly occurs in the 4th or 5th generation. Careful screening of the cultures revealed that in all cases, at least one sexual male was present in the preceding generation [7, 8]. This "spontaneous" male then releases enough sex inducer to turn sexual the whole next generation of *Volvox* gonidia in the culture. We determined the frequency of the appearence of these sexual individuals as $1.1 \times 10^{-4}$. Their generation is not triggered by an environmental or endogenous signal but by a genetic modification, which we call the *sex^c*-switch. A *sex^c* individual which occurs in both sexes at the same frequency will give rise to a stable mutant clone and will form sexual individuals in each generation without inducer present. (However, only the males will produce the sexual inducer.) The *sex^c*-switch behaves like a Mendelian mutation, but its frequency is 100 times higher than that of all other mutations in *Volvox*. We, therefore, have reason to assume that this hypermutability is based on a special mechanism, such as genetic rearrangement or cassette switching.

The existence of *sex^c* answers an old problem. Since only male spheroids release sex inducer, the question was how the first male is induced. Now it seems that a *Volvox* culture has just to reach the number of individuals at which it is almost certain that a *sex^c* male appears. By releasing sex inducer, this single mutant male is then able to induce and synchronize sexuality of the whole culture.

# 4.3.4 Sex Inducer

In this perspective, the sex inducer is a timing signal for the mass production of gametes. It is a glycoprotein, which belongs to the most potent biological molecules known, acting at concentrations down to $10^{-16}$ M, i. e. near the one-molecule-per-cell level [9]. Accordingly, the sex inducer is produced only in minute amounts, which makes biochemical, studies tedious. But by using an over productive male strain, we could obtain culture media which contain up to 10 µg inducer per litre batch. Because of the *sex^c* switch, it was not possible to scale up cultures beyond that volume. It was, therefore, quite a task to unravel the complete chemical structure of this sex inducer glycoprotein [10, 11].

## 4.3.4.1   Chemical Structure

The sex inducer is formed in several isoforms. The main biological active species are α- and β-inducer with apparent molecular masses of 28.5 and 27 kDa (SDS-PAGE), respectively. There are also smaller iso-inducer molecules present without or with at least strongly reduced biological activity. All isoforms have a common single 22 kDa polypeptide chain and differ only in the degree of glycosylation. The sex inducer contains N- and O-glycosidic sugar moieties. Removal of the N-glycosidically bound oligosaccharides abolishes the activity of the still O-glycosylated protein.

The amino acid sequence of the polypeptide chain (see also [12]) was deduced from the inducer cDNA (Fig. 4.3.2) (R. Gilles, unpublished

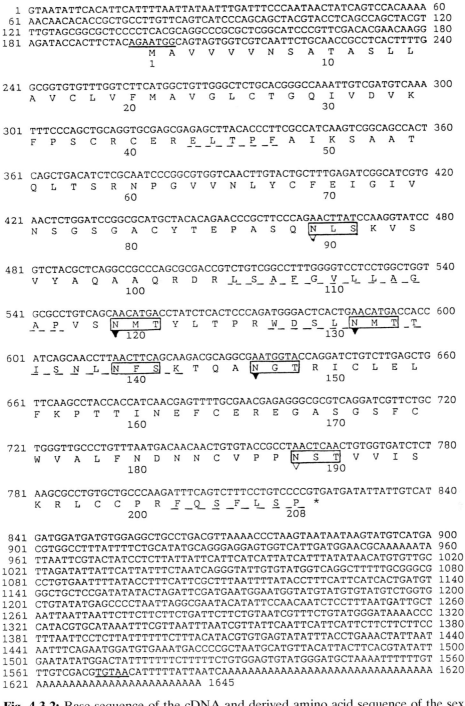

```
  1 GTAATATTCACATTCATTTTAATTATAATTTGATTTCCCAATAACTATCAGTCCACAAAA 60
 61 AACAACACACCGCTGCCTTGTTCAGTCATCCCAGCAGCTACGTACCTCAGCCAGCTACGT 120
121 TTGTAGCGGCGCTCCCCTCACGCAGGCCCGCGCTCGGCATCCCGTTCGACACGAACAAGG 180
181 AGATACCACTTCTACAGAATGGCAGTAGTGGTCGTCAATTCTGCAACCGCCTCACTTTTG 240
                   M A V V V N S A T A S L L
                   1                   10
```

```
241 GCGGTGTGTTTGGTCTTCATGGCTGTTGGGCTCTGCACGGGCCAAATTGTCGATGTCAAA 300
    A V C L V F M A V G L C T G Q I V D V K
            20                      30
```

```
301 TTTCCCAGCTGCAGGTGCGAGCGAGAGCTTACACCCTTCGCCATCAAGTCGGCAGCCACT 360
    F P S C R C E R E L T P F A I K S A A T
                    40                  50
```

```
361 CAGCTGACATCTCGCAATCCCGGCGTGGTCAACTTGTACTGCTTTGAGATCGGCATCGTG 420
    Q L T S R N P G V V N L Y C F E I G I V
                60                      70
```

```
421 AACTCTGGATCCGGCGCATGCTACACAGAACCCGCTTCCCAGAACTTATCCAAGGTATCC 480
    N S G S G A C Y T E P A S Q N L S K V S
            80                        90
```

```
481 GTCTACGCTCAGGCCGCCCAGCGCGACCGTCTGTCGGCCTTTGGGGTCCTCCTGGCTGGT 540
    V Y A Q A A Q R D R L S A F G V L L A G
                100                   110
```

```
541 GCGCCTGTCAGCAACATGACCTATCTCACTCCCAGATGGGACTCACTGAACATGACCACC 600
    A P V S N M T Y L T P R W D S L N M T T
            120                       130
```

```
601 ATCAGCAACCTTAACTTCAGCAAGACGCAGGCGAATGGTACCAGGATCTGTCTTGAGCTG 660
    I S N L N F S K T Q A N G T R I C L E L
                140                   150
```

```
661 TTCAAGCCTACCACCATCAACGAGTTTTGCGAACGAGAGGGCGCGTCAGGATCGTTCTGC 720
    F K P T T I N E F C E R E G A S G S F C
                160                   170
```

```
721 TGGGTTGCCCTGTTTAATGACAACAACTGTGTACCGCCTAACTCAACTGTGGTGATCTCT 780
    W V A L F N D N N C V P P N S T V V I S
                180                   190
```

```
781 AAGCGCCTGTGCTGCCCAAGATTTCAGTCTTTCCTGTCCCCGTGATGATATTATTGTCAT 840
    K R L C C P R F Q S F L S P *
                200             208
```

```
 841 GATGGATGATGTGGAGGCTGCCTGACGTTAAAACCCTAAGTAATAATAAGTATGTCATGA 900
 901 CGTGGCCTTTATTTTCTGCATATGCAGGGAGGAGTGGTCATTGATGGAACGCAAAAAATA 960
 961 TTAATTCGTACTATCCTCTTATTATTCATTCATCATTATCATTTATATAACATGTGTTGC 1020
1021 TTAGATATTATTCATTATTTCTAATCAGGGTATTGTGTATGGTCAGGCTTTTTGCGGGCG 1080
1081 CCTGTGAATTTTATACCTTTCATTCGCTTTAATTTTATACCTTTCATTCATCACTGATGT 1140
1141 GGCTGCTCCGATATATACTATAGATTCGATGAATGGAATGGTATGTATGTGTATGTCTGGTG 1200
1201 CTGTATATGAGCCCCTAATTAGGCGAATACATATTCCAACAATCTCCTTTAATGATTGCT 1260
1261 AATTAATTAATTCTTCTTCTTCTGATTCTTCTGTAATCGTTTCTGTATGGGATAAAACCC 1320
1321 CATACGTGCATAAATTTCGTTAATTTAATCGTTATTCAATTCATTCATTCTTCTTCTTCC 1380
1381 TTTAATTCCTCTTATTTTTTCTTTACATACGTGTGAGTATATTTACCTGAAACTATTAAT 1440
1441 AATTTCAGAATGGATGTGAAATGACCCCGCTAATGCATGTTACATTACTTCACGTATATT 1500
1501 GAATATATGGACTATTTTTTTTCTTTTTCTGTGGAGTGTATGGGATGCTAAAATTTTTTGT 1560
1561 TTGTCGACGTGTAACATTTTTATTAATCAAAAAAAAAAAAAAAAAAAAAAAAAAAAAAAA 1620
1621 AAAAAAAAAAAAAAAAAAAAAAAAA 1645
```

**Fig. 4.3.2:** Base sequence of the cDNA and derived amino acid sequence of the sex inducer. The N-glycosylated sites are indicated by black triangles ▼ on the NXS(T) triplets (boxed). Broken lines indicate the oligopeptides sequenced from tryptic digests; initiator codon and polyadenylation signal are underlined.

results). In close cooperation with K. Palme (MPI für Züchtungsforschung, Köln), a cDNA library from sexual males was constructed in lambda gt10 for this purpose. After treatment with trifluoromethanesulfonic acid, the deglycosylated sex inducer was digested by trypsin and several of the resultant peptides were sequenced at the Institut für Biochemie der Universität zu Köln. On the basis this peptide sequences oligonucleotides were synthesized, and the cDNA library was screened. We isolated a full length 1.6 kb cDNA clone, coding for a 208 amino acid protein. It contained all sequences determined at the protein level. A β-galactosidase fusion protein constructed in a pEX 3 expression vector was recognized by anti-inducer antibodies, yet it was not biologically active.

The messenger of the inducer shows some interesting features. The agaATGg initiator codon (the first one in the sequence, underlined in Fig. 4.3.2) has an A at –3 and a G at +4 position, which are both considered to be important for translation by eukaryotic ribosomes. The single open reading frame is GC-rich, a result of the strong bias in the codon usage in *Volvox* towards C > G > T > A at the third position. The polyadenylation signal is TGTAA 19 bases upstream of the poly A (underlined in Fig. 4.3.2). The same signal was also found in the unicellular relation *Chlamydomonas* [13].

The inducer protein has no similarities to other known proteins. As expected for a secreted protein, a signal peptide exists. It contains several potential cleavage sites for the signal peptidase, but we are not sure which is used, since the N-terminal amino acid of the sex inducer could not be determined because it was blocked. There are 6 potential N-glycosylation sequences, NXS or NXT (boxed in Fig. 4.3.2), the three middle-most of which are used (black triangles in Fig. 4.3.2) as confirmed by protein sequencing.

The sugar moiety makes up about 25 % of the molecular mass of the inducer. It is composed of N-glycosidic oligosaccharides bound to asparagine and O-glycosidic oligosaccharides bound to serine or threonine (the inducer contains no hydroxyproline). Total sugar analysis gave arabinose, galactose, mannose, xylose, and N-acetyl-glucosamine. We have identified three N-glycosides which are structurally related to the high-mannose type, but containing xylose, a feature which, until now, has been found only occasionally and considered characteristic for higher plants (Fig. 4.3.3). The O-glycosidic sugars, which are more difficult to analyze, are mainly mono- or disaccharides with arabinose and/or galactose as major constituents.

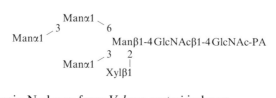

**Fig. 4.3.3:** The main N-glycan from *Volvox carteri* inducer.

All iso-inducers have a common core protein and an identical pattern of N-glycosides. They only differ in the amount of O-glycosidically bound sugar residues. Because only α- and β-inducers show maximum activity, we conclude that in addition to the protein core and the N-linked oligosaccharides a certain degree of O-glycosylation is important for the biological activity of the sex inducer. To clarify the complete role of sugar residues it will be necessary to express the inducer in a eukaryotic N-glycosylating yeast or vertebrate, if not in a *Volvox*, system.

## 4.3.4.2   Membrane Receptor

The presumed membrane receptor for the sex inducer has been looked for by the following strategy: NHS-ASA, a photoaffinity crosslinking agent, was coupled to the purified inducer protein, which was simultaneously biotinylated [11]. The modified, but still biologically active, inducer was incubated together with crude *Volvox* membranes, UV-flash fixed, and the proteins separated by SDS-PAGE. After electrotransfer, the blot was stained with avidinperoxidase. A 250 kDa protein was specifically labelled, the label of which could be quenched by the addition of unmodified inducer (Fig. 4.3.4). This was taken as evidence that the sex inducer exerts its action *via* a membrane receptor.

## 4.3.5   Inducer Action

As stated above, the inducer acts on the gonidia during embryogenesis and, from the activity titer, only a few molecules of the glycoprotein per cell suffice for switching cell cleavage. For complete sexual induction of the following generation, the inducer has to be present throughout the last 6 to 8 h before onset of the cleavages [14]. Sexual induction can be prevented by diluting the hormone or by adding inhibitors, such as anti-inducer antibodies or interfering drugs, at least 3 h before onset of embryogenesis. We, therefore, assume that this is the interval during which the pre-cleavage gonidia finally decide whether their differentiation pattern will be vegetative or sexual.

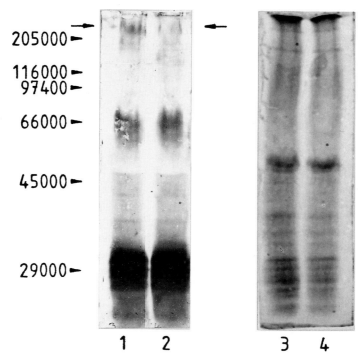

**Fig. 4.3.4:** Demonstration of a pheromone receptor on *Volvox* membranes.
1: Membrane preparation incubated with biotin-ASA inducer. 2: Same as 1 plus addition of tenfold excess unlabelled inducer; UV irradiation for 20 min.
3, 4: Coomassie Blue staining of an aliquot from 1 and 2. A specifically labelled protein band of 250 kDa is marked by arrows.

## 4.3.6   cAMP Signal Chain

One of the drugs preventing sexual induction is isobutylmethylxanthine (IBMX), a specific inhibitor of cyclic phosphodiesterase. This fact, together with the demonstration of different cAMP levels in sexual and vegetative algae, made us look for an involvement of cAMP as a second messenger of inducer action. This investigation was difficult since the usual assay systems for cAMP proved unreliable with plant material: The latter generally contains interfering substances of unknown nature. We, therefore, worked out two new assay methods for cAMP. The first was a radioimmunoassay based on a monoclonal antibody against cAMP, which was obtained in collaboration with M. Cramer (Institut für Genetik der Universität zu Köln). This RIA was highly specific and sensitive [15, 16]. The second method was based on a fluorometric determination of cAMP and other adenosine compounds after condensation with chloroacetaldehyde to the corresponding fluorescent etheno derivatives, which were measured individually after separation by HPLC [17]. Both assays gave corresponding results in *Volvox* at a detection limit in the range of 1 pmole/sample [17].

Cyclic AMP was identified in *Volvox* by both methods [18], and its kinetics followed in vegetative and induced spheroids (Fig. 4.3.5). Four minutes after the addition of the sexual inducer, the cAMP concentration peaked five-fold for 2 min. But this initial rapid inducer-specific reaction was not the only one during sexual induction. Three hours before the onset of cleavages, a second broader cAMP peak occurred exclusively in induced cultures, becoming competent to cleave sexually [19]. The time of appearence of this second cAMP peak was not determined by the time of incubation with the inducer, but by the developmental stage, namely always 3 h before the first cleavage of the gonidia. If the inducer was added too late before embryogenesis, no cAMP peak would be formed and the algae would cleave asexually. The inducer seemed to activate adenylate cyclase(s), thus raising the cAMP concentration quickly. This fast response triggered other (unknown) reactions inside the cell which ultimately led to the second cAMP peak. Without the first peak, the second peak was not triggered, or it was suppressed. It

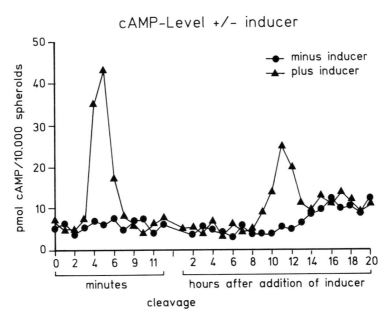

**Fig. 4.3.5:** Kinetics of cyclic Adenylate concentration in *Volvox carteri* spheroids; uninduced (●) and induced (▲).

occurred with the commitment of the gonidia to cleave and develop sexually or not.

The biphasic reaction of the cAMP level may explain the unusual features of sex induction. There has to be a fast inducer-dependent signal, since 5 to 10 min after addition of the sex inducer new induction-specific phosphorylated [20] and sulfated [21] matrix proteins appear (their significance in the induction process is still unknown). On the other hand, sexual induction remains reversible and dependent on the continuous presence of the inducer until 3 h before embryogenesis. The gonidia prepare to cleave, and then an irreversible reaction takes place which might be mediated by cAMP. We have found that inhibitors of sexual induction also inhibit the cAMP kinetics and that it is sufficient to block the second cAMP peak by $La^{3+}$ to prevent sexual induction. Thus, induction and cAMP level are causally interrelated as is to be expected if cAMP is the intracellular second message of the extracellular sex inducing signal.

## 4.3.6.1 Adenylate Cyclase and Cyclic Phosphodiesterase

Little is known to-date of the enzymes and proteins involved in plants in the cAMP signal chain. The key enzyme is adenylate cyclase (AC), which converts ATP to cAMP. *Volvox* possesses a membrane-bound AC, and N. Nass was able to purify this enzyme about hundredfold [22, 23]. A 43 kDa protein was specifically photoaffinity-labelled with 8-azido-($^{32}$P)-ATP and is weakly recognized by anti-bovine brain AC-antibodies (obtained from T. Pfeuffer, Physiol.-Chem. Institut, Universität Würzburg). There is evidence for a regulation by $Ca^{2+}$/calmodulin, and the *Volvox* enzyme behaves in this respect much like that of *Chlamydomonas*. Up to now, we have been unable to show a regulation of the AC by G-proteins, but from first data obtained in collaboration with H. Hilz (Physiol.-Chem. Institut, Universität Hamburg), we found a 40 kDa protein in *Volvox* membrane preparations which was ADP-ribosylated upon activation by pertussis toxin, which, therefore, might be a $G_1$ protein. More thorough investigation of the regulation of AC in *Volvox* is required to understand its possible role in sexual induction.

The second enzyme which can regulate the intracellular cAMP concentration is a cyclic phosphodiesterase (PDE) converting cAMP to AMP. This enzyme was also demonstrated in *Volvox* [24]. Its analysis, however, is complicated by the existence of several cAMP degrading activities, namely one cleaving cAMP to 3' AMP (its normal substrate is 2' 3' cAMP from RNA degradation), and two others involved in cAMP regulation. One of them belongs to the high and the other to the low $K_m$ class of phosphodiesterases. We could show that at the concentration of IBMX, which inhibits sexual induction *in vivo*, the low $K_m$-PDE is blocked completely *in vitro* [25].

### 4.3.6.2   Protein Kinases and cAMP Binding Protein

An early reaction to the inducer is a change in the pattern of phosphorylated extracellular proteins and the synthesis of new induction specific phospho-proteins. This points to regulated protein kinases. Our preliminary studies revealed several such activities. One at least is $Ca^{2+}$ dependent, and, we aim to show, a cAMP regulated activity.

A major stumbling block in understanding cAMP systems in plants is the failure to demonstrate cAMP binding proteins, which would be required as regulatory subunit of cAMP dependent protein kinases. O. Feldwisch has now discovered a binding protein in *Volvox* cells, which recognizes cAMP specifically with a $K_D$ of 60 nM. This 43 kDa protein has been purified to homogeneity by cAMP affinity chromatography, but its physiological function is not yet elucidated. We hope that sequencing of the protein will enable us to classify this cAMP binding protein.

## 4.3.7   Calcium and Calmodulin

Before we had direct evidence for the involvement of cAMP in sexual induction, we had hints of a possible participation of $Ca^{2+}$/calmodulin (CaM) in this process. The CaM-antagonist trifluoperazine is a potent inhibitor of the inducer action *in vivo*. If the extracellular $Ca^{2+}$ concentration, which is 0.5 mM in the standard medium, was raised to 5 to 10 mM, sexual induction is blocked. Of ions known to compete with $Ca^{2+}$, $La^{3+}$ was the most potent inhibitor [26]. At 60 μM, $La^{3+}$ can prevent induction completely without affecting viability. Since $La^{3+}$ is generally considered not to enter the cell, but to block $Ca^{2+}$ channels in the plasma membrane, we conclude that such channels, or other ion channels controlled by $Ca^{2+}$, might be opened during sexual induction. On the other hand, known calcium channel blockers, such as Verapamil and Nifedipin, though highly toxic, do not inhibit induction, possibly because they are specific for types of channels which are not involved in sexual induction.

An important indication of a functional connection between $Ca^{2+}$ channels and the cAMP system is the effect of lanthanum on the first induction-specific cAMP peak. This cAMP reaction is completely blocked in the presence of 60 μM $La^{3+}$. We conclude that the $Ca^{2+}$ dependent action precedes the cAMP-dependent reaction.

Two further findings, originally reported by R. C. Starr (Austin, TX), seem to point in the same direction [27]: On adding inducer to *Volvox* spheroids in standard medium, the normal phototaxis is blocked for about 1 h; if added to spheroids in a $Ca^{2+}$ depleted medium in which they normally can survive for several hours, the algae are immediately killed. Both effects may be explained as being due to the opening of $Ca^{2+}$ channels by the sex inducer. In the presence of the pheromone, the algae will die because they loose their intracellular $Ca^{2+}$ in the $Ca^{2+}$ depleted medium; and the photo-reactive signal will be interfered with because it uses the same signal chain ($Ca^{2+}$ fluxes) across the plasma membrane.

These experiments although suggestive, do not yet prove the involvement of calcium channels in sexual induction. This will have to be shown by direct electro-physiological methods measuring the membrane potential and calcium fluxes as effects of the action of the sexual inducer.

## 4.3.8   Conclusion

When we started with the study of cell differentiation in *Volvox*, we thought that we had chosen a system to which we could ask simple questions to get simple answers, since morphologically the cell differentiation of the algae looks so uncomplicated. But going into the molecular detail, we learned that the control of sexual induction in *Volvox* can be compared with hormone systems in higher animals and is at least as sophisticated. *Volvox* has almost the same evolutionary distance to higher plants, animals, and lower eukaryotes, including yeast and *Dictyostelium*, as revealed by srRNA sequencing [28]. In this view, it seems to be natural that *Volvox* sexual induction has some features also found in higher organisms: Transmembrane signalling *via* calcium and cAMP is clearly not restricted to higher animals, but a very "old invention" and ubiquitous in nature. Other features more closely resemble those of lower eukaryotes: The *sex*$^c$ switch obviously has similarities to the mating type switch in yeast, and the long lasting second cAMP signal reminds one of the usage of cAMP in cell differentiation of *Dictyostelium*. If *Volvox* really combines ancient and highly evolved properties of cell communication, it will be very interesting and useful to completely resolve all molecular details of sexual induction in *Volvox*. This transformation is known only in its outline to date; its elucidation should allow stimulating insights into the phylogenetic development and evolution of hormonal systems.

# 5.1 Selection of the Receptor Repertoire in the Immune System

Klaus Rajewsky*

## 5.1.1 The Starting Point: T-B Cell Collaboration

At the end of the sixties, it became clear that in the induction of antibodies two types of lymphocytes cooperate, the so-called T and B cells. From the work of my own group in Köln, as well as from the experiments of Avrion Mitchison in London, a model of T-B cell cooperation had emerged, the model of the antigen bridge [1, 2]. This model was based on the observation that antibody induction against a given epitope of an antigen required the concomitant recognition of a different epitope on the same antigen – an explanation of the "carrier effect" described earlier by Ovary and Benacerraf.

I had been attracted to immunology by the flair of its basic paradigms, such as the recognition of the unknown, immunological learning (= affinity maturation of antibodies in the course of the immune response) and memory, discrimination of self from non-self, i. e. self tolerance. Perhaps these phenomena could be approached in the frame of T-B cell collaboration – this is where my project in the Sonderforschungsbereich 74 originated.

T-B cell collaboration was only the first of a multitude of cell interactions, which were discovered in the immune system. It became more and more apparent that cell interactions were the essence of immunological regulation. Niels Jerne's network hypothesis of 1974 attributed to them an autonomous and central role in the development and the regulation of the system. The hypothesis postulated that the cells in the immune system recognize each other and regulate their functional activities through direct receptor-receptor (idiotype-anti-idiotype) interactions, independent of external antigen. Idiotypes had already attracted my interest because they appeared to offer a new tool to analyze somatic antibody diversification, and I had recruited Klaus Eichmann, an expert in this field who developed his own

---

* Institut für Genetik der Universität zu Köln

group in this Sonderforschungsbereich and with whom I closely colla-
borated. Thus it was a natural development that already in the middle of the
seventies the idiotypic network, including idiotypic T-B cell interactions,
became a main subject of our research.

The development of immunological research in the frame of the Sonderforschungsbereich 74 was strongly influenced by the close interactions with
the molecular geneticists who at that time were mainly working with bacteria
and bacteriophages and were considered by many to be a bunch of pretty
exotic and arrogant know-it-alls – one of the reasons for me to join the institute in the first place. Impressed by the potency of mutant analysis in the elucidation of biological processes in bacteria, we started early on to select
mutants of lymphoid cells *in vitro* and to analyze them in molecular terms.
We decided to concentrate on mutants of immunologically important cell
surface molecules, surface-bound antibodies initially and subsequently also
histocompatibility antigens. For selection, we used fluorescence-activated
cell sorting; in 1974, the third ever commercially available fluorescence-activated cell sorting machine was established in our laboratory through the
Sonderforschungsbereich 74. The monoclonal antibody technique which
became available in 1974 as well and which we began to use early on, improved the potential of flow-cytometric cell selection considerably and made
it a key element of our work throughout and beyond the lifespan of the
Sonderforschungsbereich 74.

I would like to end this section on the origins of the immunological projects in the Sonderforschungsbereich 74 with a final and, in my opinion, crucial point. It was in the times of this Sonderforschungsbereich that, due to
the progress of molecular biology, fundamental problems of immunology
could be directly addressed in molecular terms. The environment of the Sonderforschungsbereich was essential for us in order to develop our work as
cellular immunologists. This was not so much because of the availability of
techniques (which we had almost invariably to establish ourselves anyway),
but rather because of our continuous exposure to the traditions of thought in
molecular biology, which were the living element of the Sonderforschungsbereich and at that time pretty unique to it in the Federal Republic. I draw
from this a strong vote for diversity of content and against the streamlining of
the Sonderforschungsbereich programmes.

## 5.1.2    Interaction of (Helper) T and (Antibody Producing) B Cells

### 5.1.2.1   Helper Cell Specificity

Apparently, helper cells were capable of recognizing antigen specifically. Could they discriminate antigens as well as B cells could with the help of their antibodies? This question appeared particularly important since it touched upon the problem of the T cell receptor for antigen which was unsolved. In extensive experiments, we compared cross-reactivity (including cross-tolerance) of serum albumins of various species at the level of antibodies and T helper cell activity. We found identical patterns of cross-reactivity in both cases [3]. These results showed that helper cells exhibited exquisite immunological specificity and, in addition, suggested that helper cells might use antibodies as their receptors in the same way as B cells; the latter was later found not to be the case.

### 5.1.2.2   Memory and Immunological Tolerance in T and B Cells

The antigen bridge model, resulting from the analysis of the carrier effect, implied that immunological memory was localized both in the T and the B cell compartment [1, 2]. Was this also true for immunological tolerance? Early experiments from my group had suggested that the helper system was a prime target of tolerance induction [4, 5]. We subsequently established, in accord with others, that T helper cells could be inactivated by small doses of antigen, whereas tolerance in B cells could be achieved only by high doses of antigen and in cells producing high affinity antibodies [6, 7]. Thus, helper cells appeared to play a critical role in the protection of the organism against autoimmunity – a result whose impact could only be fully appreciated later, after the discovery of somatic hypermutation of antibodies in the generation of immunological memory (see section 5.1.4.3).

### 5.1.2.3   Mechanism of T-B Cell Interaction

The antigen bridge model postulated that the interacting cells possess receptors for distinct epitopes of the antigen bridging the two cells. In the case of the B cell, the receptor was the antibody expressed by the cell. Which signal is given by the T cell to the B cell with which it interacts? At the time there was much speculation about possible transfer of genetic information between cells of the immune system. Did the helper cell give such information to the B cell? My first Ph. D. student, Volker Schirrmacher, addressed this problem by *in vivo* experiments, using hapten-carrier systems. He showed that at least the *class* of the antibody is not determined by such an information transfer from T to B cell [8]. We decided not to follow this subject further. In the course of time, it became clear that in T-B cell interaction, T cells control B cell proliferation and differentiation, and that they do so through a variety of cytokines, which they secrete. We have shown that under certain experimental conditions cytokines can substitute for T cells in an antibody response [9]. Indeed, cytokines even play a crucial role in inducing the switch of antibody class in B cells (an effect which we used later in the analysis of class switching; see section 5.1.4.2) and they may also be involved in the hypermutation process, which operates in the pathway of memory cell generation (see section 5.1.4.3).

# 5.1.3   Network Regulation

## 5.1.3.1   Idiotypic Research Gets Underway

Since I had read Jacques Oudin's classical papers in the Journal of Experimental Medicine in 1969, I kept thinking about idiotypes as markers of antibody variable regions and their potential usefulness in the analysis of somatic antibody diversification. It was fortunate that in 1971 Klaus Eichmann, who had worked on the inheritance of idiotypes at the Rockefeller University, decided to join us and to set up a small independent research group in the Sonderforschungsbereich 74. Over the following five years, until Eichmann left Köln in 1976 to take a position at the Deutsches Krebsforschungszentrum in Heidelberg, a close contact and collaboration between his and my group developed. Eichmann successfully used idiotypes for V region gene mapping in the mouse and started on regulatory experiments using anti-idiotypic antibodies to control idiotype expression. I joined the latter area in order to explore the somatic diversification of antibodies and, in the context of my interest in T cell recognition, was curious as to whether antibody idiotypes as V gene markers would be detectable on T cells. Thus, in 1974, when Jerne put forward his hypothesis of the idiotypic network, we found ourselves already deeply involved in network experiments.

## 5.1.3.2 Idiotype Suppression and the Problem of Network-Controlled Tolerance

Anti-idiotypic antibodies turned out to be extraordinarily potent regulators of immune reactions. Eichmann found in his system of idiotypically characterized anti-streptococcal carbohydrate antibodies [10] that minute doses of xenogeneic anti-idiotypic antibodies were able to enhance or suppress idiotype expression, depending on their class [11]. One form of chronic idiotype suppression could be transferred by T cells, in fact, very small numbers of them [12]. Later, in the times of monoclonal antibodies, we showed that murine monoclonal anti-idiotypic antibodies, so-called anti-idiotopes, had similar regulatory effects in the mouse, although in our experimental system, we were never able to demonstrate idiotypic suppressor T cells as nicely, and the mode of regulation appeared to depend on antibody dose rather than class [13–16]. We also found that syngeneic anti-idiotype antibodies were potent in suppressing the *in vitro* generation of idiotype expressing cells from bone marrow pre B cells [17]. In the newborn mouse, two forms of idiotype suppression could be induced, acute and chronic suppression. Chronic suppression exhibited a particular idiotypic specificity and appeared to be controlled indirectly through T cell mediated network interactions [18, 19]. Chronic idiotype suppression could be induced by antigen as well as by anti-idiotypic antibody and exhibited immunological memory [20]. Beyond being involved in the control of the immune response, we speculated that this type of idiotype suppression could contribute to the maintenance of self tolerance [19]. I still consider this a valid notion and would like to see it tested in a direct fashion – which is not an easy task.

## 5.1.3.3 Idiotype Stimulation and Anti-Idiotypic Vaccination

Perhaps the most dramatic result of our regulatory experiments with anti-idiotypic antibodies was the phenomenon of idiotypic stimulation. In 1975, Eichmann and I published a paper in which we showed that nanogram doses of guinea pig anti-idiotypic antibody of the IgG1 class efficiently induced idiotypic memory in both B and T helper cells of mice, in the absence of any adjuvant [21]. Sam Black and Günter Hämmerling, two postdoctoral fellows in the lab, substantiated this effect by *in vitro* and genetic studies [22, 23]. Later, we showed again that idiotypic memory could also be induced by nanogram doses of syngeneic, monoclonal anti-idiotope antibodies and thus, seemed to reflect a physiological mechanism [13–16]. We suggested at the time [21] that anti-idiotypic antibodies could perhaps be used in the future as a new type of (highly standardized) vaccine. This concept has been widely taken up, and a variety of attempts have been made to develop anti-idiotypic vaccines. We would not have profited from a success in this direction: In the early times of the Sonderforschungsbereich 74 patenting scientific discoveries was not yet common practice in the scientific community.

The regulatory experiments described in sections 5.1.3.2 and 5.1.3.3 and others [24– 26] were carried out by several collaborators. I want specifically to mention my Japanese postdocs Toshitada Takemori, Takashi Saito, Shin-Ichi Nishikawa and Takeshi Tokuhisa, all of whom by now are active professors and/or group leaders in Japan, Garnett Kelsoe from the United States, Hans Tesch, a German M. D., and my students Christa Müller-Sieburg and Michael Reth.

## 5.1.3.4   Unsuccessful Attempts to Solve the T Cell Receptor Problem

The finding (mentioned above) that antibodies against an antibody idiotype (i. e. an antibody variable region) induced memory for the corresponding antigen not only in the B but also the T helper cell compartment suggested that antibodies and T cell receptors are idiotypically related and perhaps encoded by the same genes. The latter notion was further supported by experiments done in collaboration between Eichmann's and my group in which we showed that the idiotypes on both antibodies and helper cells were under the control of genes in the heavy chain linkage group [23]. Similar results were obtained by Binz and Wigzell in Sweden. A picture emerged in which T cell receptors shared heavy, but not light, chain variable regions with antibodies. The serological analysis of antigen-binding material isolated from specifically primed T helper cells by adsorption to antigen-coated nylon discs also supported this picture [27, 28].

The above experiments were based on a combination of serology and classical genetics. Although this kind of approach has been successful in many instances, it is indirect by nature and its results, suggestive as they may be, remain inconclusive. When, with the help of cloned T cells and advanced techniques of molecular biology, the T cell receptor was later isolated and its genes cloned by others, the receptor turned out to be evolutionarily and structurally related to immunoglobulin, but distinct from it. How can the discrepancy between this result and the idiotypic data be explained? The only possibility I can see (although we tried our best to exclude it [29]) is to assume idiotypic mimicry and adaptation of T and B cell receptors in the immune system on the basis of the idiotypic network [30]. However, ever since those times, I carry a deep suspicion about serological data with me and feel safer with bands on gels than antibody titers.

## 5.1.3.5 "Direct" Idiotypic T-B Cell Interaction

In the frame of the network hypothesis T-B cell interaction through antigen bridging was no more than a special case of lymphocyte interactions, which in general would be mediated by interactions of complementary receptors. Klaus Eichmann and I embarked on experiments aimed at demonstrating the latter type of T-B cell collaboration and succeeded to do so *in vitro* but not *in vivo* [31]. Indeed, a few years later, using antigen-specific, cloned helper T cells, my group provided evidence that antigen-specific T helper cells are sufficient to drive a T cell dependent antibody response *in vivo* [32, 33]. Thus, although idiotype specific helper cells can be induced *in vivo* [34] and one can easily isolate from the mouse self-reactive T cells able to activate syngenic B cells as we [35] and others have shown, the role of idiotypic T-B cell interactions in the physiology of the immune system remains elusive and a challenge for those interested in immunological regulation.

## 5.1.3.6  Structural Analysis of the Idiotypic Network

In the course of the work on idiotypic regulation, the emphasis of the general approach of the group shifted more and more from the cellular to the molecular level. This was accompanied by attempts to analyze idiotypic interactions in molecular terms. Our work in this area concerning idiotype structures, the target of idiotype suppression, the genetic basis of idiotype, and the role of somatic mutation in network interactions is described further below in section 5.1.4.3.

For a more recent review of the work described in section 5.1.3 see Ref. [36].

# 5.1.4   B Cell Differentiation and the Selection of the Antibody Repertoire

## 5.1.4.1   Early Differentiation Processes

### 5.1.4.1.1  Variable Region Gene Rearrangements During the *in vitro* Generation of B Cells from Stem Cells

In collaboration with A. Radbruch, Nobuaki Yoshida, a postdoctoral fellow from Japan, studied gene rearrangements in an *in vitro* system, in which clones of B progenitor cells developed and pre-B and B cells were generated in the course of clonal expansion [37]. He established a method by which small numbers of cells could be successfully fused to nonproducer myeloma cells. Through the analysis of the resulting hybridomas he demonstrated that the programme of gene rearrangements observed in Abelson virus-transformed B lineage cells also applied, in a first approximation, to the development of normal B lymphocytes.

### 5.1.4.1.2 Accessory Molecules are Required for Surface IgM Expression in B Cells

A crucial event in the course of B cell maturation is the expression of IgM on the surface of the cells. This marks the moment from which the cell begins contacting its antigenic environment. In collaboration with Michael Reth's group in the institute, my Ph. D. student Joachim Hombach observed by accident that upon transfection of the appropriate antibody genes into myeloma cells membrane-bound IgM was expressed in the endoplasmatic reticulum of the cells but not at their surface [38]. Through the isolation of surface IgM-positive cell variants by A. Radbruch, this led to the discovery of a protein, called B34, which associates with the membrane form of the µ chain and is required for its expression at the cell surface [39]. An area of active research has developed from these initial findings; several groups all over the world including that of M. Reth (now at the Max-Planck-Institut für Immunbiologie in Freiburg) are in the process of identifying a set of proteins, which together with the various antibody isotypes form antibody receptor complexes at the B cell surface – in striking analogy to what was already known for the T cell receptor. An understanding of the B cell response to antigens in the environment will depend on the molecular analysis of the structure and (signalling) function of this multicomponent receptor. This applies to both the induction of tolerance and that of antibody responses.

## 5.1.4.2 Antibody Class Switching

### 5.1.4.2.1 Class Switch Variants of Myeloma and Hybridoma Cells

The switch of antibody class in the immune response is a fundamental aspect of B cell differentiation, accompanying affinity maturation and the generation of immunological memory, and reflecting one of two types of DNA rearrangement typical for B cell differentiation. We first approached the analysis of the class switch by isolating class switch variants of myeloma and hybridoma cells through fluorescence activated cell sorting [40–44]. To my knowledge these were the first somatic cell mutants isolated with the help of flow cytometry – the first of many to come. The class switch variants were initially characterized at the protein level (in collaboration with Konrad Beyreuther's group); later, we used recombinant DNA techniques. The experiments told us that class switching occurred in myeloma and hybridoma cells at low frequency. It followed the order of the heavy chain constant region genes ($C_H$ genes) on the chromosome and involved recombination and deletion of DNA as had been postulated by others on the basis of the analysis of myeloma cells as such. Perhaps the most important results of these early cell sorting experiments were i) that somatic mutants could be successfully isolated if they existed in the cell population at low frequency ($10^{-5}$–$10^{-6}$) and ii) that we were in the position to isolate families of antibodies differing in antibody class but sharing identical V regions. We made use of this possibility (later also using a simplified isolation procedure which did not require flow cytometry [44]) in order to assess the role of antibody class in idiotypic regulation [16], in determining complement fixing properties [45], and in determining antibody half-lives in the serum of mice [46]. By the same approach, class switch variants were also isolated and used by several other laboratories. Koprowski's group at the Wistar Institute showed in collaboration with us that murine monoclonal antibodies of the IgG2a class were particularly efficient in eliminating human tumor cells by antibody dependent cell cytotoxicity [47]. Andreas Radbruch, my former Ph.D. student and now head of an independent research group in the department, has reviewed the subject in the Handbook of Experimental Immunology [48].

5.1.4.2.2 Mechanism of Physiological Class Switching

The cell sorting unit in the laboratory was systematically developed by Andreas Radbruch. Besides somatic cell genetics, he was interested in using fluorescence-activated cell sorting for the molecular analysis of class switching in normal (i. e. non-transformed) B-lymphocytes. Together with my Ph.D. student Fred Sablitzky he was able to show that class switching in polyclonally activated B cells was accompanied by deletion of the previously expressed $C_H$ gene [49]. In addition, when class switching to IgG1 expression was induced in such cells by interleukin 4 (IL-4), deletion occurred in both expressed and non-functional heavy chain loci, i. e. class switching did not occur randomly but was directed to a particular switch region once the cell had received the appropriate signal ("directed class switching"; [50]). This work in which another Ph.D. student, Werner Müller, was also involved, was continued by Radbruch in his own research group in the frame of the Institut für Genetik in Köln, supported by the Bundesministerium für Forschung und Technologie.

### 5.1.4.3 Antibody V Regions: Selection and Somatic Diversification

#### 5.1.4.3.1 Time for a New Approach

In the early eighties, immunologists had at their disposal all the methods required for a direct analysis of somatic antibody diversification, a central problem also of our own research. In Köln, we worked on a simple model system, which had been established by Olli Mäkelä and Thereza Imanishi in Helsinki, the antibody response of C57BL/6 mice to the hapten NP (4-hydroxy-3-nitro-phenylacetyl). This response consisted of a set of related λ chain bearing antibodies [51]. We were familiar with the monoclonal antibody technique, which allowed us to isolate antibody producing cells from different stages of the immune response. We also made a concerted effort to establish in the laboratory a spectrum of methods of molecular biology. Of particular importance at the time was mRNA sequencing, which we learned from Cesar Milstein's group in Cambridge. We also profited enormously from a collaboration with Alfred Bothwell and David Baltimore at MIT. With the new tools and on the basis of the work of Susumu Tonegawa, which had led to an understanding of the molecular mechanism involved in the generation of antibody diversity, we set out to study the selection and diversification of antibodies in the immune system and particularly in the course of the immune response.

#### 5.1.4.3.2 Molecular Structure of an Antibody Family Produced in Response to a Simple Hapten

We were the first to show in our model system that the response against a simple antigenic determinant consists of a family of closely related antibodies, which differ from each other in the fine specificity of hapten binding and idiotypic profiles. This was shown by my Ph. D. student Michael Reth through the isolation of monoclonal anti-NP antibodies from the primary (1°) and secondary (2°) response [52] and of monoclonal anti-idiotope antibodies [53]. In the 2° response, additional antibody families appeared and the response was shifted to a higher affinity for the hapten. We initiated a collaboration with D. Baltimore's laboratory in order to define antibody families at the molecular level. It turned out that the antibodies within the family expressed the same $V_L$ and $V_H$ genes but differed from each other at the $V_L J_L$ and $V_H D J_H$ borders [54, 55]. Strikingly, an antibody of this family isolated from the 2° response carried multiple somatic point mutations in both the $V_H$ and $V_L$ region [54, 56] – the first such observation in an induced immune response. This led us to a systematic analysis of the role of somatic mutation in the selection of the antibody repertoire (see below).

### 5.1.4.3.3 Somatic Hypermutation of Antibodies: Model Experiments *in vitro*

In order to explore structure-function relationships in somatic antibody mutants and their genetic basis, we isolated spontaneous V region mutants from transformed antibody producing cells *in vitro*. The cell line used in these experiments was an IgD expressing class switch variant of a hybridoma expressing an NP-specific V region [42]. V region mutants were isolated by fluorescence activated cell sorting, using pairs of monoclonal anti-idiotope antibodies and selecting for cells which had lost reactivity with one of the two anti-idiotopes [57–59]. Mutants of this kind existed in the cell population at frequencies in the order of $10^{-7}$–$10^{-8}$. In two cases, a point mutation in the V region had occurred. In a third case, the mutant phenotype was due to gene conversion, whereby a segment of the $V_H$ region had been exchanged against the corresponding segment of a neighboring $V_H$ gene [60, 61].

It is not yet clear to what extent gene conversion reflects a mechanism acting in normal B cell development. While conversions play a major role in generating antibody diversity in the chicken, they seem to be rare in the mouse, and point mutations are the major source of V gene modification in the course of the immune response [62]. In terms of function, the mutants we had isolated demonstrated strikingly that a single point mutation can completely abolish hapten binding [59] and/or idiotypic specificity [58] and can also endow the antibody with a new antigen binding specificity [59].

### 5.1.4.3.4 Relevance of Structural Data for Idiotypic Regulation

The somatic antibody mutants described in the previous section were important for our understanding of the determinants mediating idiotypic recognition. We complemented these experiments by attempts to identify in our model system the target structure of chronic idiotypic suppression. This turned out to be a peptide in CDR3 of the heavy chain [55]. We speculated that this peptide may be recognized by regulatory T cells but have not so far been able to prove this. Finally, we showed that some of the idiotypic determinants, which we had identified in the family of anti-NP antibodies, were restricted to this family, whereas others extended beyond it and were encoded by a variety of antibody structural genes [63].

### 5.1.4.3.5 Somatic Hypermutation of Antibodies: Pattern of Mutations and its Implications

Our finding of multiple somatic point mutations in an anti-NP antibody isolated from the 2° response [54] turned out to be typical of 2° response antibodies in general, including the secondary anti-NP response [64]. We also found anti-idiotypic antibodies raised by multiple immunizations to be heavily mutated ([65], for review see Refs. [62, 66]). The mutations included base substitutions in positions which do not effect the amino acid sequence of the antibody and, therefore, cannot be subject to selection by antigen. This indicates a very high rate of somatic mutation in the cells, in the order of $1 \times 10^{-3}$ per cell per generation [67] a value first suggested by M. Weigert and colleagues on the basis of more complicated considerations. The latter group had first shown that one can isolate clonally related somatic antibody mutants from individual mice undergoing an immune response and that these mutants can be ordered into genealogical trees. This indicated that the mutations were introduced into the cells in distinct mutagenic episodes in the course of clonal expansion. Shortly afterwards, Fred Sablitzky established a similar genealogical tree of somatic antibody mutants generated in an anti-idiotypic response [68] and proved the clonal relationship of the cells beyond doubt [69]. This work led us to an analysis of the role of somatic mutation in affinity maturation (see below).

### 5.1.4.3.6 Somatic Hypermutation of Antibodies: Role in Affinity Maturation

We found a particular point mutation almost invariably expressed in the $V_H$ regions of 2° response anti-NP antibodies. Using site-specific mutagenesis and gene transfection experiments, we showed directly that only this mutation was responsible for the affinity increase seen in anti-NP antibodies of the 2° over those of the 1° response. Variability in $V_H D J_H$ joints and other point mutations in the antibodies had little or no impact on affinity maturation in this system [70]. In the case of an anti-protein response, where a larger surface of the antibody is presumably involved in the interaction with the antigen than in anti-hapten antibodies, we observed stepwise affinity maturation through somatic hypermutation. Christine Kocks, a Ph. D. student, reconstructed the genealogical tree established by Fred Sablitzky (see 5.1.4.3.5) experimentally and demonstrated affinity maturation from a germ line encoded antibody specificity in at least two steps of mutant selection [71]. For this work she won the Heinz Meier-Leibnitz award of the Bundesministerium für Bildung und Wissenschaft.

### 5.1.4.3.7 Somatic Hypermutation of Antibodies: Restriction to the Pathway of Memory Cell Generation

In contrast to 2° response antibodies, antibodies of the 1° response are in general unmutated, also in our model system of the anti-NP response [55]. In what was called the 3-women-paper in the laboratory, Miriam Siekevitz, Renate Dildrop, and Christine Kocks showed that in the (adoptive) 2° response, the responding, somatically mutated cells expand without the introduction of further somatic mutations [72]. Thus, the hypermutation

analysis of lymphocytes in vitro. II. Genetic control of T-helper cell responsiveness to anti-idiotypic antibody. J. Exp. Med. 143 (1976) 861.

[24] T. Takemori, H. Tesch, M. Reth, K. Rajewsky: The immune response against anti-idiotope antibodies. I. Induction of idiotope-bearing antibodies and analysis of the idiotope repertoire. Eur. J. Immunol. 12 (1982) 1040.

[25] H. Tesch, T. Takemori, K. Rajewsky: The immune response against anti-idiotope antibodies. II. The induction of antibodies bearing the target idiotope (ab3β) depends on the frequency of the corresponding B cells. Eur. J. Immunol. 13 (1983) 726.

[26] T. Saito, T. Tokuhisa, K. Rajewsky: Induction of chronic idiotype suppression by ligands binding to the variable (not the constant) region of the idiotypic target. Eur. J. Immunol. 16 (1986) 1419.

[27] U. Krawinkel, M. Cramer, C. Berek, G. J. Hämmerling, S. J. Black, K. Rajewsky, K. Eichmann: On the structure of the T cell receptor for antigen. Cold Spring Harb. Symp. Quant. Biol. 41 (1977) 285.

[28] M. Cramer, U. Krawinkel, I. Melchers, T. Imanishi-Kari, Y. Ben-Neriah, D. Givol, K. Rajewsky: Isolated hapten-binding receptors of sensitized lymphocytes. IV. Expression of immunoglobulin variable regions in (4-hydroxy-3-nitrophenyl)acetyl (NP)-specific receptors isolated from murine B- and T-lymphocytes. Eur. J. Immunol. 9 (1979) 332.

[29] U. Krawinkel, M. Cramer, B. Kindred, K. Rajewsky: Isolated hapten-binding receptors of sensitized lymphocytes. V. Cellular origin of receptor molecules. Eur. J. Immunol. 9 (1979) 815.

[30] K. Rajewsky: Cross-reacting idiotypes on antibodies and T cell receptors: How can the data be interpreted? Scand. J. Immunol. 18 (1983) 95.

[31] K. Eichmann, I. Falk, K. Rajewsky: Recognition of idiotypes in lymphocyte interactions. II. Antigen-independent cooperation between T and B lymphocytes that possess similar and complementary idiotypes. Eur. J. Immunol. 8 (1978) 854.

[32] H. Tesch, F. I. Smith, W. J. P. Müller-Hermes, K. Rajewsky: Heterogenous and monoclonal helper T cells induce similar anti-(4-hydroxy-3-nitrophenyl)acetyl (NP) antibody populations in the primary adoptive response. I. Isotype distribution. Eur. J. Immunol. 14 (1984) 188.

[33] F. I. Smith, H. Tesch, K. Rajewsky: Heterogenous and monoclonal helper T cells induce similar anti-(4-hydroxy-3-nitrophenyl)acetyl (NP) antibody populations in the primary adoptive response. II. Lambda light chain dominance and idiotope expression. Eur. J. Immunol. 14 (1984) 195.

[34] T. Saito, K. Rajewsky: Helper T cells reacting to idiotype on IgG but not IgM. J. Exp. Med. 162 (1985) 1399.

[35] T. Saito, K. Rajewsky: A self-Ia reactive T cell clone directly stimulates every hundredth B cell and helps antigen-specific B cell responses. Eur. J. Immunol. 15 (1985) 927.

[36] K. Rajewsky, T. Takemori: Genetics, expression, and function of idiotypes. Ann. Rev. Immunol. 1 (1983) 569.

[37] N. Yoshida, A. Radbruch, K. Rajewsky: Ig gene rearrangement and expression in the progeny of B-cell progenitors in the course of clonal expansion in bone marrow cultures. EMBO J. 6 (1987) 2735.

[38] J. Hombach, F. Sablitzky, K. Rajewsky, M. Reth: Transfected plasmacytoma cells do not transport the membrane form of IgM to the cell surface. J. Exp. Med. 167 (1988) 652.

[39] J. Hombach, L. Leclercq, A. Radbruch, K. Rajewsky, M. Reth: A novel 34 kD pro-

tein co-isolated with the IgM molecule in surface IgM expressing cells. EMBO J. 7 (1988) 3451.

[40] B. Liesegang, A. Radbruch, K. Rajewsky: Isolation of myeloma variants with pre-defined variant surface immunoglobulin by cell sorting. Proc. Natl. Acad. Sci. USA 75 (1978) 3901.

[41] A. Radbruch, B. Liesegang, K. Rajewsky: Isolation of variants of the mouse myeloma X63 that express changed immunoglobulin class. Proc. Natl. Acad. Sci. USA 77 (1980) 2909.

[42] M. S. Neuberger, K. Rajewsky: Switch from hapten-specific immunoglobulin M to immunoglobulin D secretion in a hybrid mouse cell line. Proc. Natl. Acad. Sci. USA 78 (1981) 1138.

[43] H. Baumhäckel, B. Liesegang, A. Radbruch, K. Rajewsky, F. Sablitzky: Switch from NP-specific IgG3 to IgG1 in the mouse hybridoma cell line S24/63/63. J. Immunol. 128 (1982) 1227.

[44] C. E. Müller, K. Rajewsky: Isolation of immunoglobulin class switch variants from hybridoma lines secreting anti-idiotope antibodies by sequential sublining. J. Immunol. 131 (1983) 877.

[45] M. S. Neuberger, K. Rajewsky: Activation of mouse complement by monoclonal mouse antibodies. Eur. J. Immunol. 11 (1981) 1012.

[46] P. Vieira, K. Rajewsky: The half-lives of serum immunoglobulins in adult mice. Eur. J. Immunol. 18 (1988) 313.

[47] Z. Steplewski, G. Spira, M. Blaszcyk, M. D. Lubeck, A. Radbruch, H. Illges, D. Herlyn, K. Rajewsky, M. Scharff: Isolation and characterization of antimonosialoganglioside monoclonal antibody 19–9 class-switch variants. Proc. Natl. Acad. Sci. USA 82 (1985) 8653.

[48] A. Radbruch: Isotype switch variants. In: D. M. Weir, L. A. Herzenberg, C. Blackwell, L. A. Herzenberg (eds.): Handbook of Experimental Immunology, Vol. 4. Applications of Immunological Methods in Biomedical Sciences. Blackwell Scientific Publications, Edinburgh 1986, pp. 110.1– 110.12.

[49] A. Radbruch, F. Sablitzky: Deletion of $C\mu$ genes in mouse B lymphocytes upon stimulation with LPS, EMBO J. 2 (1983) 1929.

[50] A. Radbruch, W. Müller, K. Rajewsky: Class switch recombination is IgG1 specific on active and inactive IgH loci of IgG1 secreting B-cell blasts. Proc. Natl. Acad. Sci. USA 83 (1986) 3954.

[51] R. S. Jack, T. Imanishi-Kari, K. Rajewsky: Idiotypic analysis of the response of C57BL/6 mice to the (4-hydroxy-3-nitrophenyl)acetyl group. Eur. J. Immunol. 7 (1977) 559.

[52] M. Reth, G. J. Hämmerling, K. Rajewsky: Analysis of the repertoire of anti-NP antibodies in C57BL/6 mice by cell fusion. I. Characterization of antibody families in the primary and hyperimmune response. Eur. J. Immunol. 8 (1978) 393.

[53] M. Reth, T. Imanishi-Kari, K. Rajewsky: Analysis of the repertoire of anti-(4-hydroxy-3-nitro-phenyl)acetyl (NP) antibodies in C57BL/6 mice by cell fusion. II. Characterization of idiotopes by monoclonal anti-idiotope antibodies. Eur. J. Immunol. 9 (1979) 1004.

[54] A. L. M. Bothwell, M. Paskind, M. Reth, T. Imanishi-Kari, K. Rajewsky, D. Baltimore: Heavy chain variable region contribution to the NP[b] family of antibodies: Somatic mutation evident in a $\gamma$2a variable region. Cell 24 (1981) 625.

[55] A. Cumano, K. Rajewsky: Structure of primary anti-(4-hydroxy-3-nitrophenyl)acetyl (NP) antibodies in normal and idiotypically suppressed C57BL/6 mice. Eur. J. Immunol. 15 (1985) 512.

210

[56] A. L. M. Bothwell, M. Paskind, M. Reth, T. Imanishi-Kari, K. Rajewsky, D. Baltimore: Somatic variants of murine immunoglobulin lambda light chains. Nature 298 (1982) 380.

[57] M. Brüggemann, A. Radbruch, K. Rajewsky: Immunoglobulin V region variants in hybridoma cells. I. Isolation of a variant with altered idiotypic and antigen binding specificity. EMBO J. 1 (1982) 629.

[58] A. Radbruch, S. Zaiss, C. Kappen, M. Brüggemann, K. Beyreuther, K. Rajewsky: Drastic change in idiotypic but not antigen-binding specificity of an antibody by a single amino-acid substitution. Nature 315 (1985) 506.

[59] M. Brüggemann, H.-J. Müller, C. Burger, K. Rajewsky: Idiotypic selection of an antibody mutant with changed hapten binding specificity, resulting from a point mutation in position 50 of the heavy chain. EMBO. J. 5 (1986) 1561.

[60] R. Dildrop, M. Brüggemann, A. Radbruch, K. Rajewsky, K. Beyreuther: Immunoglobulin V region variants in hybridoma cells. II. Recombination between V genes. EMBO J. 1 (1982) 635.

[61] U. Krawinkel, G. Zoebelein, M. Brüggemann, A. Radbruch, K. Rajewsky: Recombination between antibody heavy chain variable region genes: Evidence for gene conversion. Proc. Natl. Acad. Sci. USA 80 (1983) 4997.

[62] C. Kocks, K. Rajewsky: Stable expression and somatic hypermutation of antibody V regions in B cell differentiation pathways. Ann. Rev. Immunol. 7 (1989) 537.

[63] R. Dildrop, J. Bovens, M. Siekevitz, K. Beyreuther, K. Rajewsky: A V region determinant (idiotope) expressed at high frequency in B lymphocytes is encoded by a large set of antibody structural genes. EMBO J. 3 (1984) 517.

[64] A. Cumano, K. Rajewsky: Clonal recruitment and somatic mutation in the generation of immunological memory to the hapten NP. EMBO J. 5 (1986) 2459.

[65] F. Sablitzky, K. Rajewsky: Molecular basis of an isogeneic anti-idiotypic response. EMBO J. 3 (1984) 3005.

[66] K. Rajewsky, I. Förster, A. Cumano: Evolutionary and somatic selection of the antibody repertoire in the mouse. Science 238 (1987) 1088.

[67] D. Allen, A. Cumano, R. Dildrop, C. Kocks, K. Rajewsky, N. Rajewsky, J. Roes, F. Sablitzky, M. Siekevitz: Timing, genetic requirements and functional consequences of somatic hypermutation during B-cell development. Immunol. Rev. 96 (1987) 5.

[68] F. Sablitzky, G. Wildner, K. Rajewsky: Somatic mutation and clonal expansion of B cells in an antigen-driven immune response. EMBO J. 4 (1985) 345.

[69] F. Sablitzky, D. Weisbaum, K. Rajewsky: Sequence analysis of non-expressed immunoglobulin heavy chain loci in clonally related, somatically mutated hybridoma cells. EMBO J. 4 (1985) 3435.

[70] D. Allen, T. Simon, F. Sablitzky, K. Rajewsky, A. Cumano: Antibody engineering for the analysis of affinity maturation of an anti-hapten response. EMBO J. 7 (1988) 1995.

[71] C. Kocks, K. Rajewsky: Stepwise intraclonal maturation of antibody affinity through somatic hypermutation. Proc. Natl. Acad. Sci. USA 85 (1988) 8206.

[72] M. Siekevitz, C. Kocks, K. Rajewsky, R. Dildrop: Analysis of somatic mutation and class switching in naive and memory B cells generating adoptive primary and secondary response. Cell 48 (1987) 757.

[73] K. Rajewsky, H. Gu, P. Vieira, I. Förster: Growth and selection of B cells in vivo. Cold Spring Harb. Symp. Quant. Biol. 54 (1989) 208.

[74] I. Förster, K. Rajewsky: Expansion and functional activity of Ly-1[+] B cells upon transfer of peritoneal cells into allotype-congenic, newborn mice. Eur. J. Immunol. 17 (1987) 521.

*Klaus Rajewsky*

[75] I. Förster, H. Gu, K. Rajewsky: Germline antibody V regions as determinants of clonal persistence and malignant growth in the B cell compartment. EMBO J. 7 (1988) 3693.

[76] H. Lemke, G. J. Hämmerling, C. Höhmann, K. Rajewsky: Hybrid cell lines secreting monoclonal antibody specific for major histocompatibility antigens of the mouse. Nature 271 (1978) 249.

[77] J. F. Kearney, A. Radbruch, B. Liesegang, K. Rajewsky: A new mouse myeloma cell line that has lost immunoglobulin expression but permits the construction of antibody secreting hybrid cell lines. J. Immunol. 123 (1979) 1548.

[78] B. Holtkamp, M. Cramer, H. Lemke, K. Rajewsky: Isolation of a clonal cell line expressing variant $H-2K^k$ using fluorescence-activated cell sorting. Nature 289 (1981) 66.

[79] B. Holtkamp, M. Cramer, K. Rajewsky: Somatic variation of $H-2K^k$ expression and stucture in a T-cell lymphoma: Instability, stabilization, high production and structural mutation. EMBO J. 2 (1983) 1983.

[80] G. Karmann, K. T. Beyreuther, M. Cramer, B. Holtkamp, S. Proska, K. Rajewsky: A structural somatic variant for the $K^k$ antigen is generated by point mutation. Immunogenetics 22 (1985) 35.

[81] R. Dildrop: A new classification of mouse $V_H$ sequences. Immunol. Today 5 (1984) 85.

[82] R. Dildrop: Classification of mouse $V_H$ sequences. In: D. M. Weir, L. A. Herzenberg, C. Blackwell, L. A. Herzenberg (eds.): Handbook of Experimental Immunology, Vol. 3. Genetics and Molecular Immunology. Blackwell Scientific Publications, Edinburgh 1986, pp. 90.1–90.6.

[83] R. Dildrop, U. Krawinkel, E. Winter, K. Rajewsky: $V_H$ gene expression in murine lipopolysaccharide blasts distributes over the nine known $V_H$ gene groups and may be random. Eur. J. Immunol. 15 (1985) 1154.

[84] R. Dildrop, A. Gause, W. Müller, K. Rajewsky: A new V gene expressed in lambda-2 light chains of the mouse. Eur. J. Immunol. 17 (1987) 731.

[85] T. Takemori, K. Rajewsky: Lambda chain expression at different stages of ontogeny in C57BL/6, BALB/c and SJL mice. Eur. J. Immunol. 11 (1981) 618.

[86] K. F. Lindahl, W. Gilbert, K. Rajewsky: Parabiosis as a model system for network interactions. In: C. M. Steinberg, I. Lefkovitz (eds.): The Immune System, Vol. 2. S. Karger, Basel 1981, p. 24.

## Doctoral Students

Volker Schirrmacher (1967–1970): Untersuchungen zur Kooperationshypothese der Antikörperinduktion: Analyse der sekundären Anti-Hapten-Antwort unter Verwendung von Sulfanyl-gekuppelten Proteinen als Hapten-Trägerkomplexe. Professor at the Institut für Immunologie und Genetik, Deutsches Krebsforschungszentrum, Heidelberg, FRG.

Dieter Armerding (1968–1971): Genetische Faktoren in der Immunantwort von Ratten. Sandoz, Wien, Austria.

Uwe Würzburg (1969–1973): Einfluß genetischer Faktoren auf die Immunantwort von Inzuchtratten. Merck, Darmstadt, FRG.

Matthias Cramer (1971–1974): Zur Genetik der monoklonalen Immunantwort von Inzuchtmäusen gegen Streptokokken-Zellwandpolysaccharide. Akademischer Rat at the Institut für Genetik, Universität Köln, FRG.

## 5.1 Selection of the Receptor Repertoire in the Immune System

Jürgen Pünter (1971–1974): Frequenz und Charakteristica von Antigen-bindenden Lymphocyten. Hoechst, Frankfurt/Main, FRG.

Inga Melchers (1971–1975): "Immune Response"-Gene der Maus: Funktionelle und genetische Aspekte der Immunantwort gegen Lactatdehydrogenase. Group leader, Research Group for Rheumatology, Universität Freiburg, FRG.

Margaret Annette Davies (1973–1975): The relationship between tumor-specific antigens of methylcholanthrene-induced sarcomata and developmental antigens on the teratocarcinoma OTT 6050.

Ulrich Krawinkel (1973–1977): Spezifische Anreicherung und Charakterisierung antigenbindender Rezeptoren von B- und T-Lymphozyten aus Mäusen und Kaninchen. Professor, Universität Konstanz, FRG.

Claudia Jack (1972–1977): Genetische Studien zur Immunantwort des Mausstammes Balb/c gegen das Zellwandpolysaccharid der Gruppe A Streptokokken. Group leader at the Deutsches Rheumaforschungszentrum, Berlin, FRG.

Andreas Radbruch (1976–1980): Isotypvarianten in der Myelomzellinie P3X63. Professor at the Institut für Genetik, Universität Köln, FRG.

Bernhard Liesegang (1976–1980): Fluoreszenzgesteuerte Zellsortierung zur Isolierung von Myelomzell-Varianten, die einen veränderten Antikörper herstellen.

Michael Reth (1977–1981): Charakterisierung individueller Antikörper und Idiotope des $NP^b$-Idiotyps. Group leader at the Max-Planck-Institut für Immunobiologie, Freiburg, FRG.

Rudi Grützmann (1976–1981): Vergleichende idiotypische Analyse von Rezeptoren mit Spezifität für Histokompatibilitätsantigene. Bayer, Wuppertal, FRG.

Bodo Holtkamp (1978–1982): Somatische H-2-Varianten. Biotest, Offenbach, FRG.

Christa E. Müller (1978–1983): Regulation der Expression idiotopischer Antikörper durch Isotyp-Varianten von monoklonalen anti-idiotopischen Antikörpern. Medical Biology Institute, La Jolla, USA.

Hans-Werner Vohr (1979–1983): Funktionelle Analyse somatischer H-2-Varianten. Bayer, Wuppertal, FRG.

Marianne Brüggemann (1980–1984): Immunglobulin-V-Gen-Varianten. Dept. of Pathology, University of Cambridge, UK.

Fred Sablitzky (1981–1985): Molekularbiologische Untersuchungen zur Selektion und Expression des Antikörperrepertoires. Group leader at the Max-Delbrück-Laboratorium, Köln, FRG.

Werner Müller (1980–1987): Charakterisierung von Lymphokinen, die B-Zellwachstum und -differenzierung beeinflussen. Group leader at the Institut für Genetik, Universität Köln, FRG.

Hans-Joachim Müller (1983–1987): Molekularbiologische Charakterisierung somatischer H-2-Varianten. Boehringer Mannheim, Penzberg, FRG.

Claudia Kappen (1983–1987): Untersuchungen zur zellulären Basis des B-Zell-Gedächtnisses in der Maus. Dept. of Biology, Yale University, New Haven, USA.

Sigrid Klein (1984–1988): Regulation der Ausprägung von Immunoglobulin-schwere-Ketten-Genen in aktivierten B-Zellen. Universitätsklinik Köln, FRG.

Christine Kocks (1985–1988): Schrittweise intraklonale Reifung der Antikörperaffinität durch somatische Hypermutation. Institut Pasteur, Paris, France.

Irmgard Förster (1985–1988): Untersuchung zur Charakterisierung der Ly1-B-Zell-Subpopulation. Institut für Genetik, Universität Köln, FRG.

Walter Weichel (1983–1989): Experimente zur Sekretion rekombinanter $H-2K^k$-Antigene. Bayer, Leverkusen, FRG.

*Klaus Rajewsky*

## Postdoctoral Scientists

Hermann Seiler (1970–1972)
Rüdiger Mohr (1971–1974), Hoechst, Frankfurt/Main, FRG.
Weng Yek Lee (1971–1972), Assistant Professor of the University of Manitoba, Dept. of Immunology, Winnipeg, USA.
Robert S. Jack (1972–1977), Group leader at the Institut für Genetik, Universität Köln, FRG.
Samuel J. Black (1973–1975), ILRAD, International Laboratory for Research on Animal Diseases, Nairobi, Kenya.
Joachim-Friedrich Kapp (1973–1975), Research Director, Research group for Rheumatology, Universität Freiburg, FRG.
Takeshi Matsunaga (1974–1976), Senior Investigator, University of Umea, Sweden.
Günter J. Hämmerling (1974–1979), Professor at the Institut für Immunologie und Genetik, Deutsches Krebsforschungszentrum, Heidelberg, FRG.
Foo Yew Liew (1975–1976), The Wellcome Research Laboratories, Beckenham, UK.
Miriam Segall (1975–1978), University of Minnesota, Dept. of Lab. Medicine and Pathology, Minneapolis, USA.
Jane Hewitt (1975–1977), The Wellcome Research Laboratories, Beckenham, UK.
Kirsten Fischer Lindahl (1976–1977), Howard Hughes Medical Institute, Dallas, USA.
Toshitada Takemori (1977–1984), Head of Department, N I H, Tokyo, Japan.
Michael Neuberger (1979–1980), MRC Laboratory of Molecular Biology, Cambridge, UK.
Joachim Neuerburg (1979–1981), University Hospital, Bonn, FRG.
Thereza Imanishi-Kari (1975–1981), Associate Professor, Tufts University School of Medicine, Boston, USA.
Mary White-Scharf (1978–1981), Repligen Corporation, Cambridge, MA, USA.
Hilmar Lemke (1979–1980), Institut für Biochemie, Universität Kiel, FRG.
Garnett H. Kelsoe (1979–1982), Assistant Professor, Dept. of Microbiology and Immunology, University of Maryland School of Medicine, USA.
Shin-Ichi Nishikawa (1980–1982), Professor, Kumamoto University Medical School, Kumamoto, Japan.
Frances Smith (1980–1983), The Mount Sinai Medical Center, New York, USA.
Raif Vasilov (1980–1982), Scientific-Industrial Association "BIOTECHNOLOGIA", Moscow, USSR.
Hans Tesch (1981–1984), Group leader, Universitätsklinik Köln, FRG.
Takeshi Saito (1982–1985), Professor, Center for Neurobiology and Molecular Immunology, Chiba University School of Medicine, Chiba, Japan.
Miriam Siekevitz (1982–1985), The Mount Sinai Medical Center, New York, USA.
Ana Cumano (1983–1987), Ontario Cancer Institute, Toronto, Canada.
Takeshi Tokuhisa (1983–1985), Professor, International Center for Medical Research (ICMR), Kobe University School of Medicine, Kobe, Japan.
Nobuaki Yoshida (1983–1988), National Kinki-Chuo Hospital, Osaka, Japan.
Deborah Allen (1984–1987), Imperial Cancer Research Fund Laboratories, London, UK.
Christa Burger (1984–1988), Dept. of Microbiology, The University of Texas Health Science Center, Dallas, USA.
Angela Gause-Pfreundschuh (1985–1987), Group leader, Universitätsklinik Köln, FRG.

Paulo Vieira (1985–1989), DNAX Research Institute, Palo Alto, USA.
Dieter Auch (1986–1988), Max-Planck-Institut für Immunbiologie, Freiburg, FRG.
Takayuki Sumida (1987–1988), The Second Department of Internal Medicine, School of
    Medicine Chiba University, Chiba, Japan.
Lise Leclercq (1987–1988), Max-Planck-Institut für Immunbiologie, Freiburg, FRG.
Ulla Weiss (1987 – present)
Daisuke Kitamura (1988 – present)


# Habilitations

Günter J. Hämmerling (1978), Professor at the Institut für Immunbiologie und Genetik,
    Deutsches Krebsforschungszentrum, Heidelberg, FRG.
Ulrich Krawinkel (1988), Professor, Universität Konstanz, FRG.
Andreas Radbruch (1989), Professor at the Institut für Genetik, Universität Köln,
    FRG.
Michael Reth (1989), Group leader at the Max-Planck-Institut für Immunbiologie,
    Freiburg, FRG.

# 5.2 Immunochemical Analysis of Antibody Binding Sites, Surface Immunoglobulins and Anti-Peptide Antibodies

Matthias Cramer*

## 5.2.1 Summary

The basic aim of the research described in this chapter was the elucidation of details concerning proteinchemical interactions of molecules and cells in the immune system. This immunochemical approach focussed on the structural analysis of antibody binding sites by photoaffinity labelling, on the detailed characterization of a set of anti-peptide antibodies, on the analysis of beta-2-microglobulin association to the MHC-chains and on the study of the membrane spanning region of surface immunoglobulin by hydrophobic membrane labelling. As part of the scientific concept of the Sonderforschungsbereich 74 this immunochemical and proteinchemical work complemented the molecular biological approaches employed by the other groups.

## 5.2.2 Structural Analysis of the Antibody Binding Site by Photoaffinity Labelling

After the establishment of all technical details of the methodology, the antigen (hapten) binding site of the murine anti-NP/NIP antibody B1–8 was studied by photoaffinity labelling (B1– 8 is analyzed in great detail by molecular biological methods in the group of K. Rajewsky, compare his report in this volume). To this end we chemically synthesized the bifunctional reagent NIP-NAP. In this molecule, NIP (4-hydroxy-5-iodo-3-nitro-phenylacetyl) is the hapten bound by the B1–8 binding site and NAP (4-azido-2-nitro-phenyl) is a photoactivatable group that crosslinks irreversibly upon illumination with visible light. After reacting B1–8 with NIP-NAP, we could show

---

\* Institut für Genetik der Universität zu Köln

via sequencing acid hydrolysis and CNBr fragments of the H-chain that lysine residues in positions 59 and 65 are labelled almost exclusively by the photoaffinity reagent. Both of these residues are located in the second hypervariable region of the B1–8 H-chain. Our results exclude the assumption that reagents like NIP-NAP (potential length: 24 Å) may have a considerable degree of flexibility once bound to an antibody binding site. The findings infer a very rigid positioning of the hapten NIP-NAP in the B1–8 binding site [1, 2].

## 5.2.3 The Use of Anti-Peptide Antibodies in the Study of Alzheimer's Disease

A series of anti-peptide antisera was prepared in the course of our studies on the molecular basis of Alzheimer's disease (AD) (compare reports by Müller-Hill and Beyreuther in this volume). AD is characterized by a massive loss of neurons in certain regions of the brain and by proteinaceous deposits in just these regions. Müller-Hill's and Beyreuther's groups were able to isolate and characterize the gene coding for the precursor protein from which the deposits in the AD brain are derived. Our set of rabbit antisera is directed against synthetic peptides designed from both the material found in the AD deposits as well as from the cDNA-derived precursor sequence. They invariably show exquisite specificity for their respective peptides. Some of them were used to demonstrate the presence of the AD-precursor protein in muscle cells. It is anticipated that these reagents will also be useful to study the proteolytic pathway(s) which lead to a regulated breakdown of the AD-precursor in healthy individuals, but to the formation of the deleterious deposits in AD-brains [3–5].

## 5.2.4 The Association of Beta-2-Microglobulin to MHC Antigens

The 12 kD protein beta-2-microglobulin exists in two forms. In one state it is bound to the cell surface-associated MHC class I chains (H-2 in the mouse, HLA in man) in a noncovalent fashion. The other state is a soluble serum protein. We have isolated and characterized a monoclonal antibody (COB6–3) that turned out to react with a combination of a single allelic product of the mouse H-2, namely H-2D$^b$, and the bovine form of beta-2-microglobulin exchanged to the murine cell culture cells from the fetal calf serum used to complement the medium. Much to our surprise, this very antibody reacts to all HLA antigens combined with human beta-2-microglobulin. We used this system to demonstrate that beta-2-microglobulin has a tendency to form dimers in solution and that the simple model of a one to one stoichiometry of MHC class I chains and beta-2-microglobulin in H-2 or HLA molecules may thus have its pitfalls [6–11].

## 5.2.5    Anchoring Surface Immunoglobulin in the Membrane

We developed a way to synthesize and purify NAP-tyramine, a substance used for hydrophobic membrane labelling. Upon addition to a cell, NAP-tyramine dissolves within the hydrophobic interior of the cellular membranes and will form covalent bonds with membrane lipids and membrane proteins once illuminated with strong visible light (compare 5.2.2). Using a surface immunoglobulin (sIg)-only transfectant developed in the groups of K. Rajewsky (compare this volume) and M. Reth we were able to analyze the membrane-spanning region of sIg by this method. Cells were labelled with NAP-tyramine and the hapten-specific sIg molecules were isolated from cell lysates. After appropriate proteolytic digestion the membrane-spanning peptides were sequenced for NAP-tyramine containing residues. Our data suggest that the intra-membrane parts of sIg possess a defined conformation and structure since they are not freely accessible to the membrane labelling reagent in all their parts [12, 13].

## 5.2.6    Cooperations

As pointed out in the introduction to this article the immunochemical approach of our group prompted numerous cooperative efforts with a number of the groups within the Sonderforschungsbereich 74, in Köln and elsewhere. It may be this aspect that reflects the enormous potential of an institution like the Sonderforschungsbereich 74 used to be in the most optimal way: such an effort does not sum up activities, it potentiates them! [14–19].

# 5.2.7 References

[1] J. Bovens: Primärstrukturen und Topologie von Immunglobulin V-Regionen. PhD Thesis, Universität Köln 1986.

[2] A. Schmitz: Molekulare Wechselwirkungen am Modell der Antikörper-Erkennung von Haptenen. Diploma Thesis, Universität Köln 1987.

[3] K. Zimmermann, T. Herget, M. J. Salbaum, W. Schubert, C. Hilbich, M. Cramer, C. L. Masters, G. Multhaup, J. Kang, H.-G. Lemaire, K. T. Beyreuther, A. Starzinski-Powitz: Localization of the putative precursor of Alzheimer's disease-specific amyloid at nuclear envelopes of adult human muscle. EMBO J. 7 (1988) 367–372.

[4] C. L. Masters, R. Martins, G. Simms, B. Rumble, S. Fuller, L. Hutchinson, J. Beer, C. Hilbich, T. Dyrks, P. Fischer, A. Weidemann, U. Mönning, G. Multhaup, M. Cramer, M. J. Salbaum, S. Wahr, K. T. Beyreuther: The molecular basis of cerebral amyloidosis in Alzheimer's disease and the unconventional virus diseases. In: A. Pouplard-Barthelaix, J. Emile, Y. Christen (eds.): Immunology and Alzheimer's Disease. Springer-Verlag, Berlin 1988, pp. 88–95.

[5] U. Mönning: Serologische Analyse von natürlichem und synthetischem Amyloid-Protein des Morbus Alzheimer. Diploma Thesis, Universität Köln 1987.

[6] B. Holtkamp, M. Cramer, K. Rajewsky: Somatic variation of H-2K$^k$ expression and structure in a T-cell-lymphoma: Instability, stabilization, high production and structural mutation. EMBO J. 2 (1983) 1943–1951.

[7] R. Mierau, M. Cramer: Serological and immunochemical analysis of H-2 class I molecules encoded by the D$^b$ region. Immunogenetics 20 (1984) 341–345.

[8] G. Karmann, K. T. Beyreuther, M. Cramer, B. Holtkamp, S. Proska, K. Rajewsky: A structural somatic variant of the K$^k$ antigen is generated by point mutation. Immunogenetics 22 (1985) 35–48.

[9] R. Mierau, P. J. Robinson, A. R. Sanderson, E. Genth, M. Cramer: Antigenic determinants shared between HLA-A, -B, -C antigens and H-2 class I molecules modified by beta-2-microglobulin. Immunogenetics 26 (1987) 351–355.

[10] M. Cramer, R. Mierau, W. Kuon, E. Weiss, P. J. Robinson: A monoclonal antibody induced by H-2 syngeneic ConA blasts: Its reactivity pattern with mouse and human MHC class I antigens. In: P. Ivanyi (ed.): MHC+X, Complex formation and antibody induction. Springer-Verlag, Berlin 1988, pp. 107–113.

[11] R. Mierau: Monoklonale Antikörper gegen Lymphozyten allotypcongener Mäuse. PhD Thesis, Universität Köln 1983.

[12] B. Bieseler: Struktur-Funktions-Analyse des Cotransportmodells Lactose-Permease. PhD Thesis, Universität Köln 1985.

[13] S. Brands: Strukturelle Analyse der Transmembran-Region von μ-Antikörper-Ketten. Diploma Thesis, Universität Köln 1989.

[14] G. Suzuki, M. Cramer, K. Hayakawa, K. Okumura, T. Tada: Idiotypic and fine specificity analysis of a 4-hydroxy-3-nitro-phenylacetyl (NP)-specific suppressor T cell hybridoma at the level of cell surface structures, isolated receptor material and functional suppressor factors. Eur. J. Immunol. 13 (1983) 711–719.

[15] R. E. Cone, R. W. Rosenstein, C. A. Janeway, G. M. Iverson, J. H. Murray, H. Cantor, M. Fresno, J. A. Mattingly, M. Cramer, U. Krawinkel, H. Wigzell, H. Binz, H. Frischknecht, W. Ptak, R. K. Gershon: Affinity purified antigen-specific products produced by T cells share epitopes recognized by heterologous

antisera raised against several different antigen-specific products from T cells. Cell. Immunol. 82 (1983) 232–245.

[16] C. Colling: Immunologische Analyse extrazellulärer Komponenten von *Volvox carteri*. Diploma Thesis, Universität Köln 1986.

[17] C. Colling, R. Gilles, M. Cramer, N. Naß, R. Moka, L. Jaenicke: Measurement of 3′:5′ cyclic AMP in biological samples using a specific monoclonal antibody. Second Messenger Phosphoprot. Res. 12 (1988) 123–133.

[18] W. Schubert, K. Zimmermann, M. Cramer, A. Starzinski-Powitz: Lymphocyte antigen Leu 19 as a molecular marker of regeneration in human skeletal muscle. Proc. Natl. Acad. Sci. USA 86 (1989) 307–311.

[19] U. Stochaj, J. Dieckhoff, J. Mollenhauer, M. Cramer, H. G. Mannherz: Evidence for the direct interaction of chicken gizzard 5′-nucleotidase with laminin and fibronectin. Biochim. Biophys. Acta 992 (1989) 385–392.

## Further References

U. Beer: Untersuchungen von Veränderungen in MHC Klasse I-Genen von Mäusefibroblasten nach Transformation mit Rous-Sarkom-Virus. Diploma Thesis, Universität Köln 1988.

M. Cramer: Isolated, hapten-specific T cell receptor material, a final view. In: J. J. Marchalonis (ed.): Antigen specific T cell receptors and factors. CRC-Press, Boca Raton 1987, Vol. II, pp. 33–54.

M. Cramer: Erkennen im Immunsystem. In: F. Cramer (ed.): Erkennen als geistiger und molekularer Prozeß. VCH, Weinheim 1991, 147–160.

S. Guder: Molekularbiologische Charakterisierung von Hodgkin-Zellinien. Diploma Thesis, Universität Köln 1988.

M. Knobloch: Zur Funktion des CD8-Antigens während der T-Zellaktivierung. Diploma Thesis, Universität Köln 1989.

B. Schneider: Antikörper gegen T-Zellen. Diploma Thesis, Universität Köln 1985.

T. Simon: Der Nachweis von Immunglobulin mRNA in Lymphozyten durch in situ Hybridisierung unter Verwendung biotin-markierter Nukleotide. Diploma Thesis, Universität Bonn 1985.

G. Strauß: HLA-transgene Mäuse als Modell für die Bedeutung von HLA-Antigenen für die Tumorabstoßung. Diploma Thesis, Universität Köln 1989.

# 5.3 Germinal Centres and the Maturation of the Immune Response

Claudia Berek*

## 5.3.1 The Immune Response to 2-Phenyl-oxazolone

Over the course of the last few years, I have studied the development of antibody diversity during an immune response. Mice were immunized with the antigen 2-phenyl-oxazolone (phOx), and at various times, spleen cells fused and hybridoma lines secreting phOx specific antibodies were established. mRNA of these lines was isolated, and the primary structure of antibody H- and L-chains was determined by directly sequencing the mRNA. It was found that in the early primary response, seven days after immunisation with antigen [1], the majority of the B cell lines expressed one particular H/L-chain combination, $V_H$-Ox1 and $V_K$-Ox1 respectively. In all these antibodies, the $V_H$-Ox1 gene segment was joined to a D/J region of 16 amino acids. The D segment always consisted of three residues, the first amino acid an Aspartic, the third a Glycine, and only the middle residue was variable. At this early stage of the immune response, little evidence for somatic mutation was seen.

The experiment was repeated 14 days after a single injection of antigen. Once again, the majority of the antibody molecules expressed the $V_H$-Ox1/$V_K$-Ox1 combination but at this point in time every H- and L-chain showed single nucleotide exchanges [2]. All sequences were diversified by a hypermutation mechanism. In $V_H$-Ox1, the mutations were distributed rather randomly over the variable region. Nucleotide exchanges were found in complementarity regions (CDR), as well as in the framework residues (FRW). A different picture was found for the $V_K$-Ox1 L-chain where mutations clustered at the end of the CDRI. In six out of eight sequences, mutations were found in residue 34 of CDRI and residue 36 of FRWII. At position 34, the germline residue Histidine was changed to either Asparagine or to Glutamine. With chain recombination experi-

---

* Institut für Genetik der Universität zu Köln

ments, it was shown that these exchanges increased the affinity about ten fold.

Further studies of the secondary [3] and the tertiary response [4] to the antigen phOx have shown that also in the memory response, the majority of the high affinity antibodies have a $V_K$-Ox1 L-chain. Practically all of these L-chains have the characteristic mutations at positions 34 and 36. However, at these later stages of the response, many more additional mutations at various positions were seen. The affinity of these antibodies was increased further [5].

The data suggested that a hypermutation mechanism is activated only after activation of B cells by antigen. This mechanism introduces nucleotide exchanges into the variable regions of the H- and L-chains with a rate of up to $10^{-3}$/base pair/generation. The re-occurrence of specific mutations which increase the affinity for antigen showed that B cell clones which carry these mutations expand preferentially and develop into the memory B cell population.

## 5.3.2 Does the Micro-Environment of the Germinal Centre Play a Role in the Maturation of the Immune Response?

We would like to know what the requirements are for the hypermutation mechanism to become operative and how it is ensured that only the high affinity variants of the original B cell clones further differentiate and develop into memory and plasma cells. It was suggested that the microenvironment of the germinal centres may play a crucial role in the maturation of the immune response [6]. Germinal centres are foci of B cell proliferation, which develop in the follicles of the lymphatic organs during the first week after immunisation. These proliferating B cells, in contrast to resting B cells, bind strongly to an agglutinin isolated from peanuts (PNA) – a marker which can be used to isolate germinal centre B cells.

## 5.3.3  Analysis of Germinal Centre B Cells

### 5.3.3.1  Isolation of PNA$^{hi}$ and PNA$^{lo}$ B Cells

Mice were immunized with the antigen phOx and at various times spleen cells were prepared from individual mice. Cells were stained with an antibody specific for the B cell marker B-220 and in addition with PNA. Using a fluorescence activated cell sorter (FACS 440), such cells could be sorted into a B cell subset that was PNA dull (PNA$^{lo}$) and a subset that was brightly stained by PNA (PNA$^{hi}$). The latter cells are the B cells, which are seen in the germinal centres.

### 5.3.3.2  Fusions of PNA$^{hi}$ and PNA$^{lo}$ Cells

Isolated B cell subsets were fused to the non-secretor line X63/Ag8.653, and hybridoma lines with specificity for the antigen phOx were isolated. Table 5.3.1 shows the results from fusions done between day 7 and day 14. Antigen-specific lines were obtained from both B cell subsets. However, the frequency of antigen specific hybridoma lines was much higher for PNA$^{hi}$ (48%) than for PNA$^{lo}$ B cells (12%). These data suggest that the small fraction of PNA$^{hi}$ B cells harbour the antigen-activated subset of B cells.

Whereas the PNA$^{lo}$ cells yielded practically only IgM secreting hybridoma lines, many IgG lines were recovered from the PNA$^{hi}$ B cells. We would like to know whether only those B cells which settle in the follicles switch from IgM to IgG.

**Table 5.3.1:** Hybridoma lines derived from PNA$^{hi}$ and PNA$^{lo}$ B cells.

|  |  | day 7 | day 8 | day 10 | day 12 | day 14 |
|---|---|---|---|---|---|---|
| PNA$^{lo}$ | IgM | 5 $(70)^2$ | 14 $(53)^2$ | 8 $(95)^4$ | 8 $(85)^1$ | 2 $(18)^1$ |
|  | IgG | – | – | – | – | 1 |
| PNA$^{hi}$ | IgM | 2 $(9)^2$ | 3 $(23)^2$ | 21 $(109)^4$ | 17 $(76)^1$ | – $(3)^1$ |
|  | IgG | 3 | 3 | 41 | 14 | 2 |

## 5.3.3.3 Antibody Diversity in Germinal Centre B Cells

Ten days after immunisation with antigen, PNA[hi] B cells were sorted on the FACS 440, fused, and the antibody diversity of phOx specific hybridoma lines analyzed. The structure of antibodies was determined by directly sequencing the mRNA of H- and L-chains. The germ line V-genes utilized by these hybridoma lines derived from germinal centre B cells and their rearrangements to D and J segments reflect what has been seen before in fusions from total spleen cells.

Most of the antibodies (13 out of 17) contained somatic mutations. In sharp contrast to the situation at a later time, when mutations are predominantly found in antibodies of the IgG class, as many mutations were seen in IgM as in IgG antibodies, suggesting that class switch and hypermutation take place at about the same time.

The distribution of somatic mutations in the sequence derived from germinal centre cells ten days after immunisation was quite different from the pattern of the immune response seen later. For instance, the typical substitutions in the $V_K$-Ox1 L-chain at codon 34, His –>Asn or –>Gln, and at position 36 from Tyr –> Phe, which were found in most of the mutated $V_K$-Ox1 L-chains, were not seen in any of the hybridomas lines derived from PNA[hi] B cells. In these sequences, 60% silent mutations were found, which is far more than one would expect from a random distribution (25%) of nucleotide substitutions. This accumulation of silent mutations suggests that many of the replacement mutations sufficiently disrupt the antibody molecule to destroy its structure or at least its ability to bind to the antigen. Cells expressing such variant receptor molecules can no longer be activated by antigen. Hence a high percentage of the antigen activated cells will die rather than differentiate further into memory and or plasma cells.

In order to test whether antigen binding loss variants can be recovered by fusion of germinal centre B cells, we screened non-binding hybridoma lines for the expression of $V_K$-Ox L-chains by the polymerase chain reaction (PCR). Out of five lines tested, one line expressed the typical $V_H$-Ox1/$V_K$-Ox1 combination. In addition, the D/J region of the H-chain had the amino acids and the length characteristic for phOx antibodies. In this line only 20% of the mutations were silent. One of the replacement mutations changed a presumptive contact residue in CDRIII of the $V_K$-Ox1 L-chain. We assume that this mutation has changed the binding site in such a way that this antibody can no longer recognize the antigen phOx.

## 5.3.4    Conclusions

Our data suggest that the small fraction of B cells which are PNA$^{hi}$ harbors the antigen activated population. B cells expressing V(D)J combinations characteristic for phOx specific antibodies are selected, presumably by antigen, to proliferate in the follicles of the lymphatic tissues, which leads to the formation of the germinal centres. Ten days after immunisation the majority of hybridoma lines derived from the PNA$^{hi}$ subset of B cells express mutated antibody molecules. However, the mutations which are characteristic for the high affinity antibodies were not observed in any of these lines. From this, we conclude that by day 10 the hypermutation mechanism is activated but that the selective expansion of those B cell clones characteristic of the mature response cannot yet be observed. In these lines, the majority of the mutation are silent (60%), suggesting that a high frequency of the replacement mutations are deleterious for the antibody structure or the antibody binding site. Only B cells which are in an activated stage can be rescued by the fusion process. Hence, the finding of a hybridoma line, which secreted antibodies of the IgG1 class but had lost specificity for the antigen phOx through mutation, suggests that during the proliferation of the B cells in the follicles, the variable region genes of the antibody molecules are diversified.

## 5.3.5    Future

Data in the literature suggest that early in the immune response B cells expressing a broad spectrum of antigen-specific receptors are activated by antigen. The majority of these cells have a low affinity for antigen [7]. It seems that only a small fraction of B cells, those with receptors of a relatively high affinity, move into the follicles and further differentiate into memory and presumably also into plasma cells. The selection of a B cell to further mature could be solely due to the higher affinity of its receptors for the antigen. However, there is evidence that the early response may be recruited from a subset of B cells which is different from the subset which gives rise to the mature response [8]. We have now started experiments, which should show whether the B cells we see in the germinal centres are indeed derived from a distinct subset of B cells.

It may be that the hypermutation mechanism becomes activated only in those cells which move into the germinal centres. In this respect, results obtained in memory responses are of interest. From the analysis of the secondary and tertiary response to phOx, I conclude that every time when the memory B cells are re-activated by antigen the hypermutation mechanism becomes operative. However, preliminary experiments suggest that memory cells do not re-enter the follicles. Certainly, the large germinal centres seen in primary responses do not appear in a memory response. This discrepancy would be resolved were the activation of hypermutation not strictly linked to the micro-environment of the germinal centre but was rather a property of a particular subset of B cells. Each time these cells are activated to proliferate by antigen, hypermutation may occur. In this case, the hypermutation mechanism would be switched on independent of the presence of B cells in germinal centres.

## 5.3.6  References

[1]  M. Kaartinen, G. M. Griffiths, A. F. Markham, C. Milstein: mRNA sequences define an unusually restricted IgG response to 2-phenyl-oxazolone and its early diversification. Nature 304 (1983) 320–324.
[2]  G. M. Griffiths, C. Berek, M. Kaartinen, C. Milstein: Somatic mutation and the maturation of the immune response to 2-phenyl-oxazolone. Nature 312 (1984) 271–275.
[3]  C. Berek, G. M. Griffiths, C. Milstein: Molecular events during maturation of the immune response to oxazolone. Nature 316 (1985) 412–418.
[4]  C. Berek, J. M. Jarvis, C. Milstein: Activation of memory and virgin B cell clones in hyperimmune animals. Eur. J. Immunol. 17 (1987) 1121–1129.
[5]  C. Berek, C. Milstein: Mutation drift and repertoire shift in the maturation of the immune response. Immunol. Rev. 96 (1987) 23–41.
[6]  I. C. M. MacLennan, D. Gray: Antigen driven selection of virgin and memory B cells. Immunol. Rev. 91 (1986) 61–85.
[7]  J. Pelkonen, M. Kaartinen, O. Mäkelä: Quantitative representation of two germ-line V genes in the early response to 2-phenyloxazolone. Eur. J. Immunol. 16 (1986) 106–109.
[8]  P.-J. Linton, D. J. Decker, N. R. Klinman: Primary antibody forming cells and secondary B cells generated from separate precursor cell subpopulations. Cell 59 (1989) 1049–1059.

# 5.3.7    Publications

C. Berek, C. Milstein: The dynamic nature of the antibody repertoire. Immunol. Rev. 105 (1988) 5–26.

C. Berek: Molecular dissection of an antigen-specific immune response. J. Auto-immun. 2 (1989) 195–201.

M.-B. Lascombe, P. M. Alzari, G. Boulot, P. Saludjian, P. Tougard, C. Berek, S. Haba, E. M. Rosen, A. Nisonoff, R. J. Poljak: Three-dimensional structure of Fab R19.9, a monoclonal murine antibody specific for the p-azobenzene-arsonate group. Proc. Natl. Acad. Sci. USA 86 (1989) 607–611.

C. Berek, M. Apel: Somatic changes in the immune response to the hapten 2-phenyl oxazolone. Progress in Immunology VII. Springer-Verlag, Berlin-Heidelberg 1989, pp. 99–105.

M. Apel, C. Berek: Somatic mutations in antibodies expressed by germinal centre B-cells early after primary immunisation. Int. Immunol 2 (1990) 813–819.

## Diploma Thesis

Martina Alscher 1988: Selektion und Analyse klonal verwandter Hybridomalinien.

# 5.4 Genetic Basic of Germ-Line Encoded Antibody Diversity

Ulrich Krawinkel*

## 5.4.1 Generation of V-Region Gene Diversity in the Germ-Line

An important question regarding the genetic basis of antibody diversity concerns the relative contribution of germ-line encoded diversity versus that generated by somatic mutational mechanisms. Germ-line encoded diversity of antibody variable regions can be attributed to the combinatorial joining of discontinuous gene segments that are juxtaposed during the development of a B lymphocyte. In immunoglobulin (Ig) heavy chains, one of several diversity ($D_H$) segments is first joined to one of the joining ($J_H$) segments, which are clustered upstream of the gene ($C\mu$) encoding the H-chain constant region of antibodies belonging to the IgM class. Subsequent joining of one of a large number of variable ($V_H$) gene segments to the preformed $D_H J_H$-element generates a complete transcription unit for the $\mu$-chain. Once this chain is produced, the joining of $V_L$ and $J_L$ gene segments results in the production of Ig light (L) chains, and a complete IgM molecule is finally displayed on the surface of a B lymphocyte. The total number of functional V-, D-, and J-segments thus constitutes the germ-line repertoire of an organism. We have studied in the murine Igh-locus the genetic mechanisms which generate multiplicity and diversity in germline $V_H$-gene segments. Our results are reviewed in the context of data from other groups in the field in Refs. [1–3].

* Institut für Genetik der Universität zu Köln
  (at present: Fakultät für Biologie der Universität Konstanz)

228

## 5.4.1.1   Multiplicity and Diversity of $V_H$-Genes in the Mouse

In most or all higher vertebrates the structure of the Igh-locus evolved by segmental amplification and divergence of a primordial V-D-J-C unit. The Igh-locus of the mouse contains a few hundred $V_H$-gene segments, which are distributed over eleven "families". Two of these families, namely VGAM3.8 and V31, were analyzed by us in detail [4, 20]. In general, $V_H$-genes sharing greater than 80% nucleotide similarity are defined as a family, whereas sequences belonging to different families share less than 70% similarity. However, the stringency of this operational definition suffers from the existence of $V_H$-genes that share 70–80% similarity with members of other well-defined families. This suggests that the $V_H$-gene repertoire is a distorted continuum of homologous genes, and "families" were generated through the amplification of some of these genes. We used the murine J558 $V_H$-gene family as a model system to analyze the evolution of $V_H$-gene families [5]. The degree of complexity of this family varies from 15 to 100 members in Igh-loci derived from laboratory and wild mice. The BALBc derived J558 $V_H$-gene family shows the highest degree of complexity. $V_H$-gene amplification seems to have occurred here less than 1 to 2 million years ago. In other Igh-loci, the J558 $V_H$-gene family has undergone a recent contraction. In conclusion, the initial expansion of the J558 family (and of the entire Igh-V locus) was presumably selected until an acceptable level of potential $V_H$-gene diversity was reached. Subsequently, expansion and contraction of $V_H$-gene families seem to have varied statistically, but a minimum $V_H$-gene pool-size is maintained. Homologous but unequal crossovers are thought to be the predominant recombinatorial mechanism generating expansion and contraction within tandemly-linked multigene families, such as $V_H$-gene families. It has important consequences for the spatial organization of $V_H$-gene loci whether unequal recombination occurs between genes of the same or between members of different $V_H$-gene families. In the first case, expansion of the involved family in one of the two recombinant loci and contraction of the family in the other one is the result, but the map order of $V_H$-families within both loci remains unchanged. Such events are self-perpetuating because an amplified locus provides more potential targets for further recombinations, as exemplified by the frequent expansions and contractions observed in the murine J558 $V_H$-gene family [5]. Homologous unequal crossovers between repetitive sequences or between conserved sequences in members of different $V_H$-gene families, in addition to expansion and contraction, generate interspersed blocks containing members of different $V_H$-gene families. $V_H$-genes showing a structure which presumably results from a recent unequal crossover are frequently found [4, 6].

A recombinatorial mechanism different from unequal crossover also seems to contribute to the generation of $V_H$-gene segment diversity, namely "gene conversion" [7, 8]. This mechanism unidirectionally transfers genetic information from a donor into a recipient gene. Upon structural analysis of genes which had been involved in gene conversion, we found that one of the two breakpoints of recombination is always associated with a palindromic sequence [8, 9]. Supportive evidence to our conclusion that palindromes can promote eukaryotic gene conversion was later published by others [10, 11].

Frequent gene conversion in the $V_H$-multigene family should involve $V_H$-pseudogenes. The latter genes could be "functional" in that they serve as sequence donors to other $V_H$-genes, thus helping to increase $V_H$-gene diversity. One would expect, therefore, that the structural integrity of pseudogenes like the one of functional genes has been selected for during evolution. The relatively "intact" structure of $V_H$-pseudogenes in the J558-family of BALB/c mice seems to support this hypothesis [12]. However, we provided evidence that these "intact" $V_H$-pseudogenes had evolved from functional $V_H$-genes only 1 to 2 million years ago and simply did not have enough time to diverge [5], that is, their intact structure is apparently not the result of a correction mechanism like gene conversion. This finding bereaves the hypothesis that $V_H$-pseudogenes remain "functional" through involvement in gene conversion events of any experimental support.

## 5.4.1.2   Organization of the Igh V Locus in Mice

A large body of evidence suggests that the order of $V_H$-gene families in the Igh-locus has a functional meaning. For instance, $V_H$-gene rearrangements in pre B lymphocytes seem to be influenced by the chromosomal position of a $V_H$-gene. A map of $V_H$-genes is therefore required to understand phenomena such as biased $V_H$-gene usage in foetal lymphoid tissue.

The recombinatorial mechanisms discussed in the previous chapter have shaped the murine Igh-locus in a way that $V_H$-gene families are now organized as interspersed gene clusters [1–3, 13–15]. In Igh-loci, newly generated by recombination between two distinct haplotypes, restriction fragment length polymorphism (RFLP) can be used to assign $V_H$-genes to either haplotype and to order them with respect to the breakpoint of recombination [13, 14]. Most recombination breakpoints in the murine Igh-loci map to the 3' end of the V-locus, thus providing good map resolution for this recombinogenic region [14]. Our genetic map of the murine Igh-V-locus has been confirmed by "deletion" mapping, which takes advantage of the fact that V- to DJ-joining results in the deletion of the chromosomal region located between the rearranging $V_H$-gene and the

DJ-region. Therefore, the order of $V_H$-genes can be analyzed with respect to the endpoints of deletions associated with VDJ-joining [13]. Genetic and deletion analyses can resolve the relative $V_H$-gene order but provide little information on the physical distances between genes. In order to establish a physical map, we analyzed the murine Igh-locus by pulsed field gel electrophoresis and chromosomal walking utilizing lambda-phage clones. We linked the $V_H$-gene family J558 to the V31 family, and the $V_H$-families Q52, PC7183 and VGAM3.8 to the D-J-C-region of murine Igh-locus [13, 15]. The $J_H$-proximal $V_H$-gene in the mouse belongs to the VGAM3.8 family [14, 15]. We also found $D_H$-segments, which are interpersed with $V_H$-genes [15]. A synopsis of all mapping analyses of the murine Igh-locus is shown in Ref. [1].

Although there seem to be few restrictions on recombinations, in three distinct haplotypes from laboratory mice, a similar map order of $V_H$-gene families is found. Interspersed members of families PC7183, Q52, and VGAM3.8 always seem to be located at the 3' end of the Igh V-locus. A preserved map order of $V_H$-gene families could indicate selection during evolution, that is, a functional meaning to this order. A second line of evidence also suggests a functional significance of the gene order in the Igh V-locus. Members of $V_H$-gene families PC7183, Q52, and VGAM3.8 are preferentially used in cells derived from foetal lymphoid tissue. The genetic mechanism responsible for this is unknown, but seems to effect a preferential rearrangement of J-proximal $V_H$-genes (reviewed in [16]).

Considering both, a programmed rearrangement of members of J-proximal $V_H$-genes and the apparently preserved position of families PC7183, Q52, and VGAM3.8 proximal to the $J_H$-locus, one might speculate that members of these families encode antibody specificities that are needed during the early development of the immune system. Selection for the correctly timed expression of these antibody specificities could thus mean selection for a J-proximal position, that is, the preferential rearrangement, of the $V_H$-genes encoding them.

## 5.4.2    Class Switch Recombination

The analysis of the molecular basis of class-switching in B lymphocytes was performed in collaboration with A. Radbruch, Köln. Our data are reviewed in Ref. [17]. This report summarizes the data concerning the molecular analysis of switch regions in the murine Igh-locus.

Individual activated B lymphocytes can switch the heavy chain constant region of the antibody that they produce while retaining the antibody's specificity. Recombination plays an important role in the course of class-switching. The existence of switch (S) regions in front of constant (c) region genes indicates that specific recombinatorial mechanisms requiring typical sequences are involved. The differences in the structures of the various switch regions raise the possibility of class-specific switch recombination. Since the corresponding control mechanisms are largely unknown, we surveyed recombinant switch regions isolated from the genome of activated murine B lymphocytes for sequences which could have served as recognition sites for recombination enzymes. In accordance with other groups, we found that the enzymes involved in class-switching to Cγ3 and to Cγ1 do not recognize an obviously well-conserved sequence motif, although all analyzed class-switches were mediated by the structurally well-defined switch regions [18]. We now think that class-switch recombination is regulated at the level of chromatin structure, which is presumably influenced by higher-order DNA structures in the switch regions.

In a project also concerning class switch regulation, we started to analyze mRNAs that are specifically induced by lipopolysaccharide in lymphocytes. In the initial phase of this project, we developed a method of cDNA cloning [19] which, in a modified version, is now used in many research groups.

# 5.4.3   References

[1]   U. Krawinkel, T. Christoph, T. Blankenstein: Organization of the Ig $V_H$-locus in Mice and Humans. Immunol. Today 10 (1989) 339–344.
[2]   T. Blankenstein, G. Lehle, R. Schüppel, C. Kolb, C. Kappen, E. Weiler, U. Krawinkel: Content and Organization of the Immunoglobulin Heavy Chain Variable Gene Locus in the Mouse. In: Molecular Basis of the Immune Response, Annals of the New York Academy of Sciences, Vol. 545 (1988) 192–194.
[3]   M. J. Taussig, M. J. Sims, U. Krawinkel: Regulation of Immunoglobulin Gene Rearrangement and Expression. Immunol. Today 10 (1989) 143–146.
[4]   E. Winter, A. Radbruch, U. Krawinkel: Members of novel $V_H$-gene Families are found in VDJ-regions of polyclonally activated B-lymphocytes. EMBO J. 4 (1985) 2861–2867.
[5]   T. Blankenstein, F. Bonhomme, U. Krawinkel: Evolution of Pseudogenes in the Immunoglobulin $V_H$-gene Family of the Mouse. Immunogenetics 26 (1987) 237–248.
[6]   T. Blankenstein, G. Zoebelein, U. Krawinkel: Analysis of Immunoglobulin heavy chain V-region Genes belonging to the $V_{NP}$-gene Family. Nucl. Acids Res. 12 (1984) 6887–6900.
[7]   U. Krawinkel, G. Zoebelein, M. Brüggemann, A. Radbruch, K. Rajewsky: Recombination between Antibody Heavy Chain variable Region Genes. Proc. Natl. Acad. Sci. USA 80 (1983) 4997–5001.
[8]   T. Blankenstein, G. A. Rathbun, P. W. Tucker, U. Krawinkel: Recombination between $V_H$-Pseudogenes. Mol. Immunol. 26 (1989) 319–322.

[9] U. Krawinkel, G. Zoebelein, A. L. M. Bothwell: Palindromic Sequences are associated with Sites of DNA Breakage during Gene Conversion. Nucl. Acids Res. 14 (1986) 3871–3881.

[10] G. F. Hatful, S. M. Noble, N. D. F. Grindley: The gamma-delta resolvase induces an unusual structure at the recombinational crossover point. Cell 49 (1987) 103–110.

[11] D. K. Nag, M. White, T. D. Petes: Palindromic sequences in heteroduplex DNA inhibit mismatch repair in yeast. Nature 340 (1989) 318–320.

[12] J. B. Cohen, D. Givol: Conservation and divergence of immunoglobulin $V_H$-pseudogenes. EMBO J. 2 (1983) 1795–1800.

[13] T. Blankenstein, U. Krawinkel: Immunoglobulin $V_H$-region Genes of the Mouse organized in overlapping Clusters. Eur. J. Immunol. 17 (1987) 1351–1357.

[14] G. Lehle, C. Kolb, C. Kappen, R. Schüppel, E. Weiler, U. Krawinkel: A map of $V_H$-genes located next to the $D_H$-region in the Igh-locus of two congenic Igh recombinant Mouse Strains. Eur. J. Immunol. 18 (1988) 1275–1281.

[15] T. Christoph, U. Krawinkel: Physical Linkage of variable diversity and joining Gene Segments in the Immunoglobulin Heavy chain Locus of the Mouse. Eur. J. Immunol. 19 (1989) 1521–1523.

[16] F. W. Alt, K. T. Blackwell, G. D. Yancopoulos: Development of the primary antibody repertoire. Science 238 (1987) 1079–1087.

[17] U. Krawinkel, A. Radbruch: Immunoglobulin Heavy Chain Class-Switching. In: F. Calabi, M. Neuberger (eds.): Molecular Genetics of Immunoglobulin. Elsevier Science Publishers, Amsterdam 1987, pp. 135–151.

[18] E. Winter, U. Krawinkel, A. Radbruch: Directed Ig-class Switch Recombination in activated murine B-cells. EMBO J. 6 (1987) 1913.

[19] U. Krawinkel, R. Zoebelein: Rapid Synthesis of cDNA for cloning into Lambda Vectors. Nucl. Acids Res. 14 (1986) 1913.

[20] M. Sims, U. Krawinkel, M. Taussig: Characterisation of Germline Genes of the VGAM 3.8 $V_H$ Gene Family from BALB/c mice. J. Immunol. (1992) in press.

## Members of the Group

Thomas Blankenstein (1983–1988, Diploma and PhD Thesis), Institut für Immunologie, Universitätsklinikum Steglitz, Berlin, FRG.

Thomas Christoph (1988–1989, Diploma Thesis), Institut für Genetik, Universität Köln, FRG.

Ulrich Krawinkel (1981–1988, head of the group, Habilitation), Klinische Forschergruppe für Rheumatologie, Universität Freiburg, FRG.

Osman Sözeri (1987–1988, Diploma Thesis), Institut für Genetik, Universität Köln, FRG.

Ekkehard Winter (1983–1988, Diploma and PhD Thesis), Stifterverband der deutschen Industrie, Essen, FRG.

Rele Zoebelein (1981–1988, technician), Institut für Genetik, Universität Köln, FRG.

I gratefully acknowledge the commitment of R. Zoebelein to the work of our group.

*Ulrich Krawinkel*

## Guest Scientists

Georg Lehle (1987), Universität Konstanz, FRG.
Anand Raghavachar (1985), Universität Ulm, FRG.
Ralf Schüppel, Universität Konstanz, FRG.
Martin J. Sims, University of Cambridge, UK.


## Doctoral Theses

Thomas Blankenstein (1983–1987): Molekularbiologische Untersuchung zur Evolution und genomischen Anordnung von Immunglobulin $V_H$-Genen.
Ekkehard Winter (1984–1988): Molekularbiologische Untersuchungen zur Umlagerung und Aufprägung der Gene im IgH-Lokus polyklonal aktivierter B-Lymphocyten.

234

# 6    Virology

# 6.1 Malignant Transformation by Adenovirus: Integration of the Viral DNA and Control of Viral Gene Expression by Specific Patterns of DNA Methylation[1]

Walter Doerfler*

## 6.1.1 Setting the Goal

While working on the expression of N gene heteroduplex molecules of bacteriophage lambda DNA in Dave Hogness' laboratory at Stanford during the summer of 1965 [1, 2], I became interested in the problem of oncogenesis and gene regulation by mammalian DNA viruses. Human adenoviruses (Fig. 6.1.1), with a genome size not too different from that of bacteriophage lambda, appeared to be suitable tools to investigate molecular mechanisms in mammalian cells, just as bacteriophages in bacteria had served that role in the classical period of molecular biology. "Viruses as packages of genes and genetic elements with millenia of biological experience" [5] have proven their importance in the forefront of molecular biology. In 1966, I had the good fortune that Igor Tamm at The Rockefeller University offered me the opportunity for independent research in the heart of Manhattan Island. The American Cancer Society provided a grant (1969–1972) and the Health Research Council of the City of New York a Career Scientist Award (1969–1971).

During the time at Rockefeller University (1966–1971), I could demonstrate that hamster cells were abortively infected by adenovirus type 12 (Ad12) [6], that the viral DNA or fragments of Ad12 DNA became covalently linked to hamster cell DNA starting at about 16 h postinfection [7, 8], and that Ad12 DNA could not replicate in hamster cells [9]. At the same time, the viral DNA was characterized by physical-chemical methods and by electron microscopy [10]. It could also be shown that in hamster cells the parental Ad12 DNA was cleaved to distinct fragments presumably by a cellular endonuclease [11]. An endonuclease can be associated with the adenovirus penton [12], but this cell-coded enzyme could be separated

[1] This chapter is dedicated to David S. Hogness, Stanford University, Stanford, CA., Igor Tamm, The Rockefeller University, New York, N.Y., and Wolfram Zillig, MPI, Martinsried.

* Institut für Genetik der Universität zu Köln

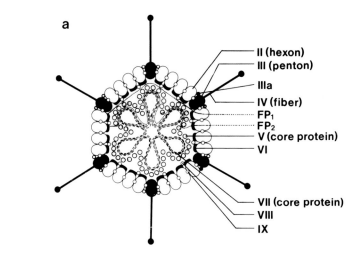

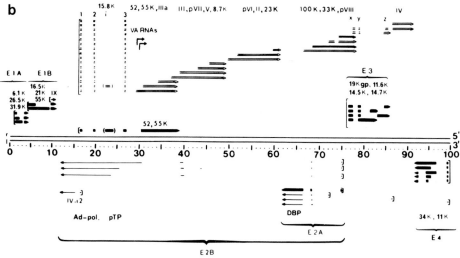

**Fig. 6.1.1:** (a) Structural model of the adenovirion. The Roman numerals refer to the standard designations of the viral structural proteins. FP stands for fracture plane. In freeze etching studies combined with electron microscopy as well as in biochemical studies on the adenovirion, the planes designated by dotted lines coincided with regions along which the structural elements of the virion could be peeled apart. This figure was taken from [3].

(b) Genetic map of adenovirus type 2. The coding capacities of individual genome segments are indicated by the sizes of polypeptides (K = 1,000 dalton) or by the designations of the virion structural proteins (Roman numerals). The double stranded DNA molecule with its 3' and 5' ends and a scale in map units is in the center of the graph. For details see text and [4].

from the penton [13]. Part of the endonuclease work was performed during a sabbatical stay in Lennart Philipson's laboratory at Wallenberg laboratoriet, Uppsala Universitet (1971/72) with a fellowship from the Swedish Cancer Society.

The finding of integrated Ad12 genomes was the first reported for oncogenic viral genomes and had interesting functional implications in that it offered an explanation for the persistence and continued expression of viral DNA in adenovirus-transformed cells or in Ad12-induced tumor cells. Moreover, the existence of integrated adenoviral genomes raised the question to what extent insertional mutagenesis could be involved in adenovirus oncogenesis [8]. This question still remains to be answered 20 years later. We started to investigate whether viral and cellular DNA were covalently linked also in productively infected cells [14].

## 6.1.2   Continuing Research in Köln

After a long search that continued well past 1972, I realized that the Institut für Genetik in Köln would be as close to a department structure as one could hope to find in Germany. Moreover, with the Sonderforschungsbereich 74 "Molecular Biology of the Cell" of the Deutsche Forschungsgemeinschaft a granting system had just been initiated in Köln, which allowed for flexible support and interesting interactions with other scientists. Willi Stoffel, Hans Eggers, and Peter Starlinger convinced me that Köln would be a good place for active research. Within the Sonderforschungsbereich 74, it was also possible to support independent group leaders who directed their own research programs. Over the years, Dennis T. Brown, Ernst-L. Winnacker, and Helmut Esche held, or most recently, Dagmar Knebel-Mörsdorf holds such positions within the Division of Virology at the Institut für Genetik. Most importantly, we have been fortunate to train and motivate a considerable number of very serious and interested students and young scientists.

At the time of my moving to Köln, I felt that a detailed study of viral DNA integration and the regulation of viral gene expression, particularly in adenovirus-transformed cells, was worth considerable effort. At about the time, Max Delbrück asked me what I was planning to work on. When I finished explaining, he dryly remarked "You will not be alone". I knew that he was right. But a rapidly expanding field with a lot of information coming in from many laboratories provided a tremendous stimulus, although one had to be skilful to avoid non-productive duplications. Traveling and the establishment of many international contacts were about the only safeguard against that undesirable possibility. In that context, it was very helpful that the Universität zu Köln and the Land Nordrhein-Westfalen provided an opportunity for one semester sabbaticals. I spent stimulating months in the Department of Biochemistry at Stanford in 1978 and in the Department of Molecular Biology at Princeton in 1986 [15, 16]. During my 1988 vacations, I worked at Kawasaki Medical School in Kurashiki, Japan [17].

In this report, I am going to summarize the results adduced on the following major projects. Details of these investigations will be found in the cited references.

– Insertional mutagenesis as a model for transformation by oncodna viruses.
– On the mechanism of integrative recombination in mammalian cells.
– The species-specificity of the major late promoter of Ad12 DNA.
– The long-term inhibition of promoter activity by sequence-specific methylation.
– Transcription of the genome of the *Autographa californica* nuclear polyhedrosis virus (baculovirus).

## 6.1.3 Insertional Mutagenesis as a Model for Transformation by Oncodna Viruses

Various possibilities exist of how adenovirus infection can lead to the malignant transformation of cells. In adenovirus-transformation, the viral functions can exert a decisive effect on cellular gene activities [18]. Excitement has also been generated by the observation that E1A-encoded proteins can complex with the product of the retinoblastoma gene [19]. Could the delicate balance between oncogene and anti-oncogene products be disturbed in this way?

In our own work, we have found that the apparently total loss of the Ad12 genome, including the E1 region, from Ad12-induced hamster tumor cells can still be compatible with the oncogenic phenotype of these cells [20]. Of course, it cannot be ruled out that small genetic elements, like enhancers of viral origin, may still have been retained in these tumor cells. These findings have alerted us to reconsider the hit-and-run mechanism of DNA virus transformation. This frequently discussed, yet unproven, mechanism is also compatible with that of insertional mutagenesis. In that model, the adenoviral DNA is thought to integrate at a number of cellular sites which are sensitive for the regulation of cellular growth control. This integration can occur without [21] (Fig. 6.1.2) or with the deletion [22] of cellular nucleotides. Thus, even after the excision of the

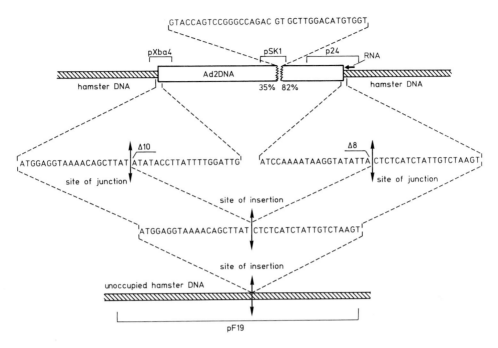

**Fig. 6.1.2:** Integrated Ad2 genome in unique hamster cell DNA in cell line HE5. The sequence data summarized here have been published elsewhere [21, 25].

adenoviral DNA from the cellular genome [23], the mutagenic effect at the site of the previous insertion of foreign DNA could be explained.

The model of insertional mutagenesis can only be seriously contemplated if adenoviral DNA insertion was not restricted to a very limited number of cellular sequences. We have, therefore, undertaken a detailed study of eight different sites at which adenovirus DNA has recombined with hamster, mouse, or human DNA. In this initial attempt, established adenovirus-transformed cell lines or Ad12-induced tumor cell lines, as clonal derivatives of original transformation events, have been selected for sequence analyses of the sites of junction between adenovirus DNA and cellular DNA [21, 22, 24–29]. Of particular interest has been the site of junction between Ad12 DNA and human DNA in the virion-encapsidated symmetric recombinant SYREC2 [30, 31]. The structure of this latter junction site has unequivocally proven recombination to occur between viral and human cellular DNA also in productively infected human cells [32–35]. There is evidence that such recombination/integration events can be selective [36] and may be preferentially targeted to chromosome 1 in Ad12-infected cells [37].

A summary of the main conclusions derived from the analytical work on sites of junction between adenovirus and cellular DNAs is presented in Table 6.1.1. A comparison of the nucleotide sequences from eight different sites of junction has not supported the possibility that adenovirus DNA inserts at specific cellular DNA sequences [27]. I consider it unlikely

**Table 6.1.1:** Integration of Adenovirus DNA into the Host Genome.

1. State of integrated viral DNA

– Dependence on permissivity of cell system. Ad12 DNA – frequently integrated intact in non-permissive hamster cells.
Ad2 DNA – internal deletions; fragments integrated.
– Site of recombination at or close to the termini of viral DNA (terminal protein; termini are recombinogenic).
– Multiple copies (1 to >30) of viral DNA per cell genome.
– Rarely true tandems of adenovirus DNA.
Interspersed cellular or rearranged viral sequences.
Amplifications of terminal sequences.
– Frequently, not invariably, deletions of terminal nucleotides (8–174 nucleotides).
– Viral DNA associated with specific chromosomes.
However, more than one site of chromosomal location.
– Loss of viral genomes observed giving rise to revertants.
Total loss or 1/2 to 1 out of 22 copies of Ad12 DNA can be preserved (cell line T637).
– Loss of viral DNA compatible with oncogenic phenotype.
– In addition to intact Ad12 genome(s), dispersed fragments of Ad12 DNA can be integrated.

2. State of cellular DNA at integration site

– Unique, specific cellular integration sequence is *not* apparent. However, the possibility exists that initially integration of foreign DNA occurs at selective sites, and that subsequently foreign DNA becomes repositioned such that integration appears to be non-specific.
– Patch homologies between viral and cellular DNAs are observed, but not always found.
– Integration into unique or repetitive cellular DNA sequences.
– Deletion of cellular DNA (1.5 to 1.6 kbp) observed or integration without the loss of a single cellular nucleotide.
– Specific genes in vicinity of integration site in some cases: 4.5S RNA, endogenous intracisternal A particle (IAP) genome.
– Transcriptional activity of cellular DNA at or close (360 nt.) to site of adenovirus DNA integration has been documented for six different preinsertion sequences.
– The results of *in vitro* recombination experiments in a cell-free system suggest that preinsertion sequences (p7, p16) can be preferred sites of recombination in comparison to randomly selected hamster DNA sequences.

that foreign (viral) DNA can integrate completely randomly, but insertion can rather take place at many different sites. One critical element in the evaluation of these results deserves to be emphasized. For technical and conceptual reasons, we had chosen clonal lines of transformed and tumor cells. Thus, the possibility still exists that at early times of adenoviral infection, a limited number of integration sites is orginally chosen. However, due to post-integrational rearrangements of the inserted adenovirus genomes, a much larger number of cellular sequences may eventually appear to have served as targets for integration. This problem requires further investigations on sites of viral DNA integration early after adenovirus infection [7, 8].

We have studied the transcriptional activity of preinsertion sites and have been able to document that all the investigated cellular preinsertion sequences in hamster, mouse, and human cells were transcriptionally active (Fig. 6.1.3) in untransformed, non-infected cells or in the DNA from hamster organs [38, 39]. It is a plausible working hypothesis that foreign DNA would preferentially integrate into transcriptionally active cellular sequences with a certain chromatin structure that might render the cellular DNA more easily accessible for the recombination with foreign DNA. Moreover, it might be advantageous for foreign genes to be inserted into actively transcribed regions of the cellular genome with regulatory sequences in the immediate vicinity. Perhaps foreign genes can thus be more immediately expressed in the "language" of the new host cell.

The integration of adenovirus DNA is a reversible process in that it has been documented in several instances that the viral DNA can be excised in part or *in toto* from the genome of the transformed cells [20, 23, 40–42]. Sometimes the loss of viral genomes is accompanied by a concomitant alteration of the cellular phenotype. The total loss of the viral genome from Ad12-induced tumor cells is still compatible with the tumorigenic phenotype [20]. At the present time, we do not understand the mechanism by which viral genomes are excised from the cellular genome. In the genome of the Ad12-transformed hamster cell line T637, about 20 genome equivalents of Ad12 DNA are integrated [43–45]. Although these genomes are located on different chromosomes in the T637 cells [46], they can apparently all be lost jointly. This finding renders the excision mechanism even more enigmatic. *In vitro* studies on this mechanism procured promising results [47].

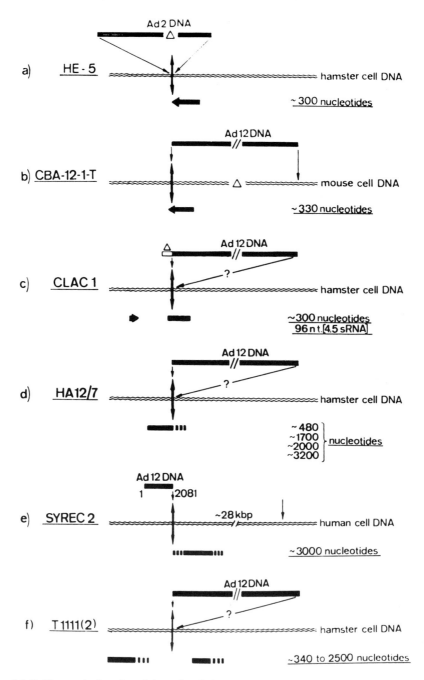

**Fig. 6.1.3:** Transcriptional activity of cellular preinsertion sites for the integration of adenovirus genomes. The viral cellular junction sequences of cell lines HE5 (Ad2-transformed hamster cells: Refs. [21, 25]), CBA-12-1-T (Ad12- induced mouse tumor cells: Ref. [22]), CLAC1 (Ad12-induced hamster tumor cells: Ref. [26]), HA12/7 (Ad12-transformed hamster cells: Ref. [29]), SYREC2 (symmetric Ad12-human cell

244

## 6.1.4  On the Mechanism of Integrative Recombination in Mammalian Cells

Judging from the results of our earlier integration studies [7, 8, 27, 32, 48; for reviews 49, 50], recombination between viral and cellular DNA appears to be a rather frequent event. It is, however, unknown how stably the integrated viral DNA remains fixed in the host genome upon repeated cell passages. The adenovirus system will then be suitable for studies on the mechanism of recombination in mammalian cells. We have devised a cell-free system using nuclear extracts from BHK21 hamster cells to follow recombination between Ad12 and hamster DNA *in vitro*. The novelty in this approach [51] lay in the selection of the cellular hamster DNA pre-insertion sequence corresponding to the integration sequence from the Ad12-induced hamster tumor CLAC1 [26] or T1111/2 [28] as one re-combination partner. These pBR322 cloned preinsertion sequences have been termed p7 or p16, respectively, and might carry motifs that have been recognized at least once by the cellular recombination system at a time before the tumors CLAC1 and T1111/2 have originated. The p7 sequence is shown in Fig. 6.1.4a. Upon *in vitro* incubation of the PstI frag-ments of Ad12 DNA and p7 or p16 DNA with nuclear extracts, recombi-nants containing Ad12 DNA have been isolated via recA⁻ strains of *Escherichia coli*. Many of these recombinants have been characterized [52] (Fig. 6.1.4b). A large number of control experiments has ascertained that recombination has taken place in the cell-free extract and not in the pro-karyotic host used for the isolation of individual recombinants [52]. At a comparable frequency, similar recombinants have not been generated, when randomly selected hamster DNA sequences of comparable lengths have been used in these experiments. Hence, there is indeed a preference for recombination of Ad12 DNA with the preinsertion sequences p7 and p16. Within the Ad12 DNA molecule, the segment between map units 60 and 70 exhibits a very marked preference for recombination with the preinsertion hamster DNA, perhaps because of structural characteristics [52]. Preliminary results of experiments in which we have tried to fraction-ate the nuclear extracts look promising. This system may enable us to elucidate at least some of the essential elements, which catalyze the inser-tion of adenoviral (foreign) DNA into the host genome.

DNA recombinant: Refs. [30, 31]), and T1111(2) (Ad12-induced hamster tumor cells: Ref. [28]) have been cloned and sequenced. The corresponding cellular preinsertion sequences are all transcriptionally active. The map locations of the RNAs synthesized and their sizes are indicated below each integration site scheme. For details, see Refs. [38, 39].

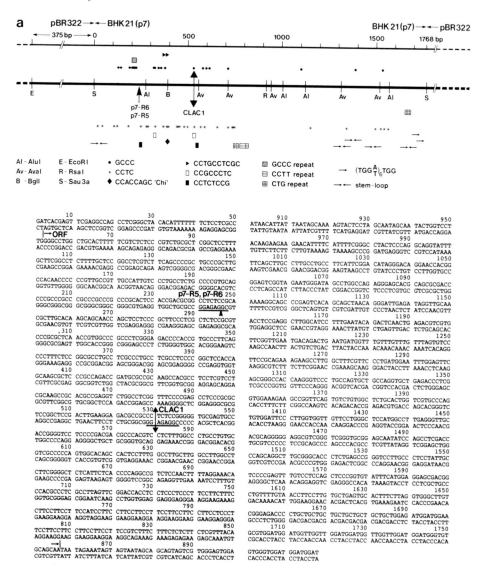

**Fig. 6.1.4:** Origin and characterization of *in vitro* generated recombinants between Ad12 DNA and the preinsertion hamster sequence p7. These recombinants were isolated after the incubation of the PstI fragments of Ad12 DNA and the pBR322-cloned hamster sequence p7 with nuclear extracts from BHK21 hamster cells [52].
(a) Restriction map and nucleotide sequence of the 1768 bp hamster DNA p7. CLAC1 ↕ designates the site of Ad12 DNA insertion in cell line CLAC1 (cf. legend to Fig. 6.1.3).
(b) Characteristics and map locations of several p7-Ad12 recombinants (p7-R1 etc). Note clustering of recombinants in the 60 to 70 map unit segment of Ad12 DNA. For details and further explanations, see Ref. [52].

246

**b**

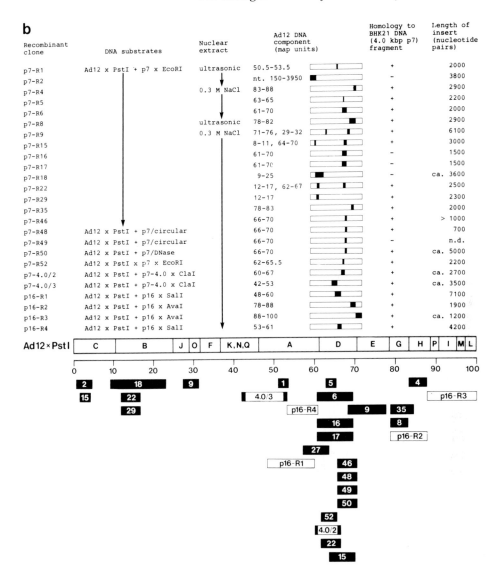

| Recombinant clone | DNA substrates | Nuclear extract | Ad12 DNA component (map units) | Homology to BHK21 DNA (4.0 kbp p7) fragment | Length of insert (nucleotide pairs) |
|---|---|---|---|---|---|
| p7-R1 | Ad12 x PstI + p7 x EcoRI | ultrasonic | 50.5-53.5 | + | 2000 |
| p7-R2 | | | nt. 150-3950 | - | 3800 |
| p7-R4 | | 0.3 M NaCl | 83-88 | + | 2900 |
| p7-R5 | | | 63-65 | + | 2200 |
| p7-R6 | | | 61-70 | + | 2000 |
| p7-R8 | | ultrasonic | 78-82 | + | 2900 |
| p7-R9 | | 0.3 M NaCl | 71-76, 29-32 | + | 6100 |
| p7-R15 | | | 8-11, 64-70 | + | 3000 |
| p7-R16 | | | 61-70 | - | 1500 |
| p7-R17 | | | 61-70 | - | 1500 |
| p7-R18 | | | 9-25 | - | ca. 3600 |
| p7-R22 | | | 12-17, 62-67 | + | 2500 |
| p7-R29 | | | 12-17 | + | 2300 |
| p7-R35 | | | 78-83 | + | 2000 |
| p7-R46 | | | 66-70 | + | > 1000 |
| p7-R48 | Ad12 x PstI + p7/circular | | 66-70 | + | 700 |
| p7-R49 | Ad12 x PstI + p7/circular | | 66-70 | - | n.d. |
| p7-R50 | Ad12 x PstI + p7/DNase | | 66-70 | + | ca. 5000 |
| p7-R52 | Ad12 x PstI x p7 x EcoRI | | 62-65.5 | + | 2200 |
| p7-4.0/2 | Ad12 x PstI + p7-4.0 x ClaI | | 60-67 | + | ca. 2700 |
| p7-4.0/3 | Ad12 x PstI + p7-4.0 x ClaI | | 42-53 | + | ca. 3500 |
| p16-R1 | Ad12 x PstI + p16 x SalI | | 48-60 | + | 7100 |
| p16-R2 | Ad12 x PstI + p16 x AvaI | | 78-88 | + | 1900 |
| p16-R3 | Ad12 x PstI + p16 x AvaI | | 88-100 | + | ca. 1200 |
| p16-R4 | Ad12 x PstI + p16 x SalI | | 53-61 | + | 4200 |

## 6.1.5 The Species-Specificity of the Major Late Promoter of Ad12 DNA

With the intent to understand mechanisms by which DNA viruses transform mammalian cells to tumor-like cells, we have undertaken a detailed study of the interactions of Ad12 with hamster cells (Table 6.1.2). Ad12

**Table 6.1.2:** The Non-Permissive Interaction of Hamster Cells with Human Adenovirus Type 12.

| | |
|---|---|
| Absence of virion production. | [6, 53, 54] |
| Absence of Ad12 DNA replication. | [6, 9, 55, 59, 60] |
| Early Ad12 genes are expressed as documented by:<br>– presence of early Ad12-specific RNA<br>– presence of mRNA translatable into early proteins: exception, 34K protein. | [56–58]<br>[61, 63] |
| Lack of expression of late Ad12 genes. | [58] |
| Deficiency of the major late promoter (MLP) of Ad12 in hamster cells. | [68] |
| Mitigator element in first intron of Ad12 MLP. | [69] |
| L1 and VA RNAs are not expressed. | [64] |
| The complementation of Ad12 deficiencies by Ad2 functions. | |
| Hamster cells double-infected by Ad2 (or Ad5) and Ad12<br>– replicate Ad12 DNA<br>– do not replicate Ad12 virions. | [60] |
| Hamster cells which carry in an integrated state and constitutively express the E1 region of Ad2 or of Ad5 DNA (plus internal segments, i. e., map units 32.4–41.4)<br>– replicate Ad12 DNA<br>– express late Ad12 genes<br>– do not synthesize late Ad12 proteins, or only in limited amounts<br>– do not replicate Ad12 virions<br>– BHK297–C131 infected with Ad12 express Ad12 L1 and VA RNAs, cell line BHK-Ad2E1 does not. | [59]<br>[60] |
| E1B functions of Ad2 or Ad5 DNA are related to complementation of Ad12, E1A functions may also be involved. | |

cannot replicate in hamster cells [6, 53, 54], and the viral replication cycle is interrupted before viral DNA replication [6, 7, 9, 55]. However, the deficiency of Ad12 in hamster cells is complex and most likely involves more than one viral function. It is also of significance for the characterization of this virus-cell system that another adenovirus type, Ad2, can efficiently replicate in hamster cells [6, 53].

## 6.1.5.1   Ad12-Specific Functions in Hamster Cells

Many of the early functions of Ad12 DNA are transcribed in hamster cells [56–60]. It is not certain whether the full complement of all the early Ad12 messenger RNAs are synthesized in hamster cells. In Ad12-transformed cells, many of the early Ad12 genes are transcribed, but selective transcriptional blocks in individual Ad12 functions have been observed. In the Ad12-transformed hamster cell line HA12/7, e.g., RNA from the early region 3 (E3) is not transcribed. It is also significant that Ad12-specific late functions fail to be transcribed after the infection of hamster cells, in many Ad12-transformed cells or in Ad12-induced tumor cells [58–60]. Of the early Ad12 messenger RNAs transcribed in hamster cells, most can be translated in a cell-free translation system [61–63]. However, a 34 kiloDalton (kD) protein of Ad12 DNA fails to be translated in these experiments. By a comparison to the early Ad12-specific proteins synthesized *in vitro* with hybrid-selected RNAs from Ad12-infected, permissive human cells, it is apparent that most other early Ad12 gene products can be synthesized *in vitro* when using RNAs from Ad12-infected hamster cells. The defect to produce the 34 kDa protein in Ad12-infected BHK21 cells might be of significance in explaining the abortive infection cycle.

In Ad12-infected hamster cells, the virus-associated (VA) RNA and RNAs encoded in the late L1 region of Ad12 DNA are not transcribed [64]. The late region L1 of adenovirus DNA is expressed already early in productive infection [65–67]. The L1 55/58 kDa gene products are required for the assembly of virus particles [15].

The major late promoter (MLP) in adenovirus DNA governs the expression of almost all the late viral functions. Since the late Ad12 functions cannot be transcribed into stable messenger RNAs in hamster cells, the activity of the MLP of Ad12 or of Ad2 DNA has been investigated in human and in hamster cells. For this purpose, the MLP of Ad12 DNA or of Ad2 DNA has been fused to the prokaryotic gene of chloramphenicol acetyltransferase (CAT). Both constructs are active in human cells. However, the Ad12 promoter does not function in hamster cells, whereas the Ad2 promoter is active in hamster cells [68]. In these experiments, adenovirus-infected cells have to be used, since the MLP is dependent on the availability of other viral and of cellular functions. Thus, the MLP of Ad12 is capable of differentiating between cellular auxiliary functions of different species. These results document that the deficiency of the Ad12 genome in hamster cells can be at least partly attributed to the species-specificity of the MLP.

## 6.1.5.2 A Mitigator Element in the Ad12 MLP

The MLP of Ad12 DNA (nucleotides –228 to +438 relative to its cap site) does not function in uninfected or in Ad12-infected hamster BHK21 cells [68]. We have demonstrated more recently that the sequence between nucleotides +249 and +438 in the Ad12 MLP, relative to nucleotide +1 as the site of transcriptional initiation, is in some way responsible for the late transcriptional block of this promoter in hamster cells [69]. An Ad12 MLP-CAT construct comprising nucleotides −228 to +248 shows striking activity in hamster cells, and its activity is very markedly enhanced in Ad2- or in Ad12-infected hamster or human cells in comparison to the nucleotide −228 to +438 construct. By using the Bal31 nuclease, a series of constructs has been generated to localize precisely the position of the promoter element, which mitigates or abolishes promoter function. The data demonstrate that this element comprises between 30 and 40 nucleotides [69]. Transcription is initiated at the authentic Ad12 MLP cap site in the nucleotide −228 to +248 construct after the transfection of both hamster and human cells. The Ad2 MLP does not seem to harbor a comparable mitigator element. These results [69] adduce evidence for the presence of a mitigator element in the first intron following the Ad12 MLP. This mitigator contributes to the abortive infection cycle that Ad12 is subject to in non-permissive hamster cells and may explain the relatively low efficiency of Ad12 in human cells.

### 6.1.5.3 Partial Complementation of Ad12 Defects in Hamster Cells by Ad2 and Ad5 Functions

The block in the replication and late expression of the Ad12 genome in hamster cells can be complemented by Ad2 (Ad5) gene products [59]. Upon the coinfection of hamster cells with Ad12 and Ad2 (or Ad5), both Ad12 DNA and Ad2 (Ad5) DNA replicate. Ad2 (Ad5) virions are assembled in doubly infected cells, the production of Ad12 particles is not detectable. For a more refined analysis, and in order to avoid difficulties with the interpretation due to even weak homologies between Ad12 and Ad2 (Ad5) DNAs, Ad2- or Ad5-transformed hamster cell lines, which contain in an integrated form and constitutively express only the left terminus (E1 region) of Ad2 or Ad5, have been superinfected with Ad12. Ad12 DNA replicates in these cell lines, and many of the early and late Ad12-specific RNAs are transcribed and stable in these hamster cells. It is unknown whether all Ad12-specific RNAs are produced in this complementing system. Ad12 virions fail to be assembled [59, 60]. In the Ad5-transformed cell line BHK-C131 [70], the Ad12 L1 and VA RNA segments are stably transcribed [64]; these functions fail to be expressed, however, in the Ad2-transformed hamster cell line BHK-Ad2E1 [71]. The possibility, therefore, exists that for the Ad12 complementation effect in Ad2- or Ad5-transformed hamster cell lines, cellular functions may be – in part – responsible [64]. We have also adduced evidence for the notion that the Ad12-complementing functions in the Ad5 genome reside predominantly but not exclusively in the E1B region of Ad5 DNA [60]. It is likely that E1A functions also have a role in the complementation process.

inhibited. It is not known how maintenance methylation is regulated, nor how frequently this mode of demethylation occurs, if it occurs at all.

b) The cytidine analogue 5-azacytidine causes demethylation of replicating DNA, presumably by the specific inhibition of the DNA methyltransferase. Consequently, previously shut-down genes are transcriptionally reactivated [92–94].

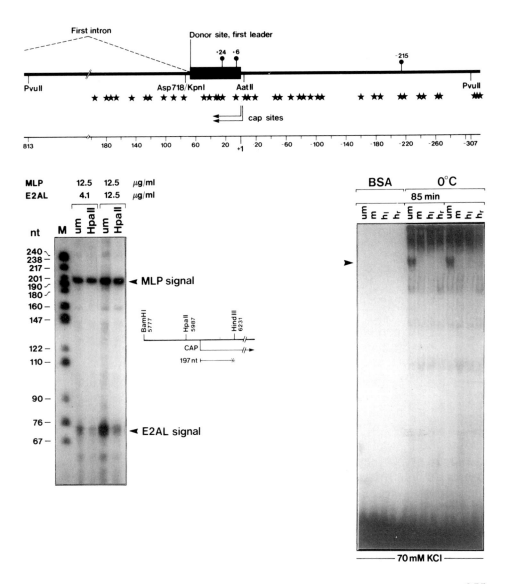

c) There are more subtle ways to reactivate a promoter which had been inhibited by sequence-specific methylations. Trans- and cis-acting factors can reactivate the methylated late E2A promoter of Ad2 DNA without demethylation to occur, at least not in both DNA complements. The 289 amino acid E1A-encoded classical transactivator of adenovirus can turn on the previously silenced late E2A promoter [71, 86] with an activity somewhat lower than that of the unmethylated promoter. This trans-effect can be elicited with the methylated late E2A promoter either chromosomally or episomally (plasmid) located and with the transactivating gene itself being localized episomally or on cellular chromosomes, respectively [71, 94]. Transcription from the trans-activated, methylated E2A promoter is also initiated at the authentic cap site. Similar observations have been made in cis-activating experiments in which the strong immediate early enhancer of human cytomegalovirus has been placed in the vicinity of the methylated late E2A promoter. This promoter responds with activity levels comparable to those of the unmethylated promoter [95]. It is thus apparent that the inhibiting effect of sequence-specific methylations on promoter activity is not unconditional, but has to be viewed in the context of other factors influencing promoter function.

## 6.1.6.4 The Spreading of DNA Methylation

Newly arising patterns of DNA methylation in mammalian DNA, particularly in DNA sequences recently integrated into that genome, are characterized by the spreading of DNA methylation from a seminal point to adjacent DNA sequences. In several hamster cell lines, an E2A promoter-CAT gene construct has been genomically fixed [89], and this construct had previously been *in vitro* methylated at three 5'-CCGG-3' sequences by HpaII DNA methyltransferase [89]. With increasing generation numbers of these established cell lines, methylation extends progressively to neighboring 5'-CG-3' nucleotide pairs, as shown by the genomic sequencing technique [96]. Eventually, all 5'-CG-3' sequences are completely methylated (Fig. 6.1.6), and methylation also involves a 5'-CA-3' and a 5'-CT-3' dinucleotide in one of the cell lines [90].

The spreading of DNA methylation at the nucleotide level is paralleled by the finding that an adenovirus genome of some 30 to 35 kilobase pairs in length, which has been integrated into the hamster cell genome, becomes also methylated progressively and non-randomly with an increasing number of cell generations [41, 42, 42a]. It is unknown whether the pre-existing patterns of methylation in the neighboring cellular sequences or particular structural arrangements of recently inserted foreign DNA determine the generation of a new pattern in the inserted viral DNA sequences. It will be interesting to investigate whether the progressive inactivation of an entire X chromosome in mammals can also be caused by the gradual extension of DNA methylation across the entire X chromosome [97]. The spreading of DNA methylation at the megabase level can be envisaged if one assumes seminal foci of methylation spaced in certain intervals on that chromosome to coalesce.

## 6.1.6.5  Sequence-Specific Methylations and DNA-Protein Interactions in Eukaryotic Promoters

The interaction of some, not all, of the numerous transcription factors with specific promoter motifs can be inhibited by DNA methylation. By applying the techniques of genomic sequencing and *in vivo* dimethylsulfate genomic footprinting, it has been demonstrated that the spreading of methylation in cell lines, which carry the late E2A promoter with three *in vitro* premethylated 5′-CCGG-3′ sequences, initially involves a DNA domain of this promoter, which is devoid of bound proteins. Subsequently, methylation further spreads to neighboring regions irrespective of the presence of proteins, and the patterns of complexed transcription factors are altered [90].

Possible explanations for the inhibitory effect of three 5-mC residues in the late E2A promoter sequence are structural alterations in DNA or the positive or negative modulation of the sequence-specific binding of proteins. A synthetic oligodeoxyribonucleotide of 50 basepairs (bp) or a restriction endonuclease fragment of 73 bp in length, which comprises the +24 and +6 5′-CCGG-3′ sequences of the late E2A promoter, has been methylated or hemimethylated at these two sites, or has been left unmethylated and has subsequently been incubated with a partly purified nuclear extract of human HeLa cells. Protein binding has been monitored by electrophoretic migration delay of the [$^{32}$P]-labeled 50 bp oligodeoxyribonucleotide or the 73 bp fragment on polyacrylamide gels. The formation of one of the DNA-protein complexes in this analysis has been compromised when 5′-CCGG-3′ methylated oligodeoxyribonucleotides have been used in the binding assays (Fig. 6.1.5) [98]. Similar results have been obtained when the 50 bp oligodeoxyribonucleotide has been hemimethylated in either complement. The formation of the same complex can also be obliterated by adding the same non-methylated oligodeoxyribonucleotide as competitor to the reaction mixture. The methylated oligodeoxyribonucleotide does not act as a competitor, nor does a randomly composed oligodeoxyribonucleotide of identical length. The results document that protein binding is abolished by methylation of the same sequences in the late E2A promoter, whose methylation inhibits promoter function. Further analytical work will be required to determine how promoter methylation inhibits promoter activity.

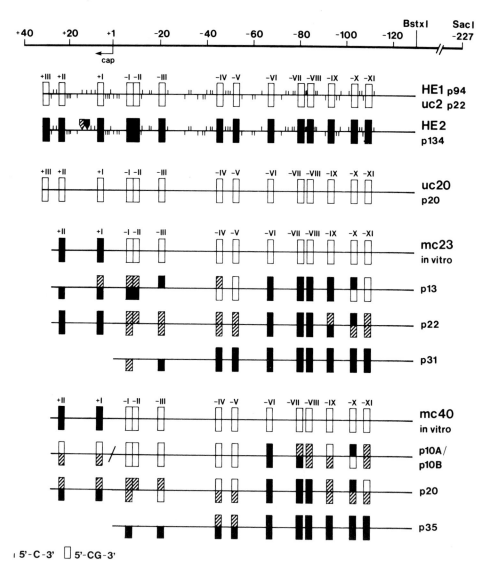

**Fig. 6.1.6:** Spreading of DNA methylation in the late E2A promoter of Ad2 DNA. Summary of the genomic sequencing data. The locations of fully and partially methylated or unmethylated 5'-CG-3' dinucleotides in the late E2A promoter of Ad2 DNA were presented for the transformed cell lines HE1, HE2, uc2, uc20, or for cell lines mc23 and mc40 in different passages (p), as shown by genomic sequencing [90, 96]. The scale refers to nucleotide numbers in the late E2A promoter relative to the cap site (◄). The 5'-CCGG-3' sequences are at nucleotides +6 (+I) and +24 (+II), which have been *in vitro* premethylated in the generation of cell lines mc23 and mc40. Horizontal lines represent the late E2A promoter segment in individual cell lines. The 5'-CG-3' sequences in this segment (+III to –XI) are represented by vertical bars: □

## 6.1.7 Transcription of the Genome of *Autographa californica* Nuclear Polyhedrosis Virus (Baculovirus)

In 1976, we started to develop a second viral system, which has proved very useful in studies on the molecular biology of insect cells. More recently, the baculovirus *Autographa californica* nuclear polyhedrosis virus (AcNPV) has become one of the very efficient eukaryotic expression vector systems [99].

Projects linked to the AcNPV work in my laboratory have been supported by the Sonderforschungsbereich 74 for only a relatively short time (1986–1988). Therefore, the major results will be summarized only briefly. After establishing the system [100, 101], we have been mainly interested in the mode of viral gene transcription in productively-infected insect cells.

a) We have mapped 11 early and some 90 late RNA size classes on the viral genome [102] and also presented a preliminary translational map [103]. RNA splicing is not a major mechanism of RNA processing in AcNPV-infected *Spodoptera frugiperda* insect cells [102]. The occurrence of small splices or splicing very late in infection cannot be ruled out.

b) In several segments of the viral genome, we discovered nested sets of overlapping RNA transcripts with common 5'- or 3'-termini [104]. Later, this mode of viral transcription was confirmed for many other parts of the AcNPV genome.

c) We have determined the nucleotide sequence in the 81.2 to 85.0 map unit fragment of AcNPV DNA, precisely mapped the nine major RNA size classes, which are transcribed from this segment, and have located five open reading frames in that segment of the AcNPV genome [105]. The findings are compatible with the notion that, by translating different RNA size-classes from that region, different polypeptides can be expressed. Perhaps the array of different RNAs of varying lengths and coding capacities provides a viable alternative to RNA splicing for the flexible expression of the viral genome [104, 105].

---

unmethylated, ■ completely methylated, and ▨ 5'-CG-3' sequences, which are methylated in only some of the integrated promoter copies. The bars above the horizontal line designate 5'-CG-3' dinucleotides in the top strand of the promoter sequence, the bars below the line represent the same dinucleotides in the bottom strand. The positions of a methylated 5'-CA-3' (▼) and a methylated 5'-CT-3' (▨) dinucleotide are also indicated [90].

d) In applied research, the baculovirus vector system has been used for the expression of several hundred eukaryotic genes in insect cells [106]. The vector utilizes the very efficient polyhedrin promoter. Thus, appreciable levels of foreign gene products can be synthesized. Secondly, *Spodoptera frugiperda* insect cells infected with recombinant AcNPV have been shown to perform most of the essential posttranslational modifications (glycosylation, cleavage, phosphorylation, nuclear or membrane transport) on the newly synthesized polypeptides and thus render them biologically functional. Lastly, baculoviruses appear to be a biologically safe vector system, since the AcNPV genome cannot replicate or be transcribed in mammalian cells, nor does the viral genome persist in mass cultures of AcNPV-inoculated mammalian cells [107].

In collaborations with the groups of Hans-Dieter Klenk (Gießen/Marburg) or of Peter Starlinger (Köln), we have expressed the hemagglutinin of fowl plague virus in insect cells [108] and in lepidopteran larvae [109], or the Activator (Ac) gene product (perhaps a transposase) of *Zea mays,* respectively [110]. The fowl plague virus hemagglutinin can be glycosylated in insect cells in a pattern very similar to that found in avian or mammalian cells [111]. The recombinant hemagglutinin is also cleaved, transported to the cell membrane, and protects chickens against viral infection [108].

In many respects, the baculovirus system has turned out to be a fascinating object for studies in molecular virology, and we will pursue studies on various aspects of the virus.

## Acknowledgements

It is a pleasure to acknowledge the commitment of and the exciting collaboration with many of our students and young scientists whose names appear on the following pages as authors of the cited publications. Without their interest and dedication, this work could not have been accomplished. Moreover, I am indebted to Petra Böhm for excellent editorial work.

This research would not have been possible without the long-standing support by the Deutsche Forschungsgemeinschaft through SFB74-C1. The projects were also financed by the basic budget (Grundausstattung) of the Universität zu Köln and by donations from Fonds der Chemischen Industrie, Frankfurt/Main. Fellowships were provided by the Alexander von Humboldt-Foundation and the Boehringer-Ingelheim-Fonds.

# 6.1.8 References

[1]     W. Doerfler, D. S. Hogness: The strands of DNA from lambda and related bacteriophages: Isolation and characterization. J. Mol. Biol. 33 (1968) 635–659.

[2]     W. Doerfler, D. S. Hogness: Gene orientation in bacteriophage lambda as determined from the genetic activities of heteroduplex DNA formed in vitro. J. Mol. Biol. 33 (1968) 661–678.

[3]     D. T. Brown, M. Westphal, B. T. Burlingham, U. Winterhoff, W. Doerfler: Structure and composition of the adenovirus type 2 core. J. Virol. 16 (1975) 366–387.

[4]     G. Akusjärvi, U. Pettersson, R. J. Roberts: Structure and function of the adenovirus-2 genome. In: W. Doerfler (ed.): Adenovirus DNA: The Viral Genome and its Expression. Martinus Nijhoff Publishing, Boston-Dordrecht-Lancaster 1986, pp. 53–95.

[5]     W. Doerfler: Genetic principles and viral oncogenesis. In: E. Grundmann (ed.): Cancer Campaign, Experimental Neurooncology, Brain Tumor and Pain Therapy. Gustav Fischer Verlag, Stuttgart-New York 1987, Vol. 10, pp. 47–55.

[6]     W. Doerfler: Nonproductive infection of baby hamster kidney cells (BHK21) with adenovirus type 12. Virology 38 (1969) 587–606.

[7]     W. Doerfler: The fate of the DNA of adenovirus type 12 in baby hamster kidney cells. Proc. Natl. Acad. Sci. USA 60 (1968) 636–643.

[8]     W. Doerfler: Integration of the deoxyribonucleic acid of adenovirus type 12 into the deoxyribonucleic acid of baby hamster kidney cells. J. Virol. 6 (1970) 652–666.

[9]     W. Doerfler, U. Lundholm: Absence of replication of the DNA of adenovirus type 12 in BHK21 cells. Virology 40 (1970) 754–757.

[10]   W. Doerfler, A. K. Kleinschmidt: Denaturation pattern of the DNA of adenovirus type 2 as determined by electron microscopy. J. Mol. Biol. 50 (1970) 579–593.

[11]   B. T. Burlingham, W. Doerfler: Three size-classes of intracellular adenovirus deoxyribonucleic acid. J. Virol. 7 (1971) 707–719.

[12]   B. T. Burlingham, W. Doerfler, U. Pettersson, L. Philipson: Adenovirus endonuclease: association with the penton of adenovirus type 2. J. Mol. Biol. 60 (1971) 45–64.

[13]   U. Reif, U. Winterhoff, U. Lundholm, L. Philipson, W. Doerfler: Purification of an endonuclease from adenovirus-infected KB cells. Eur. J. Biochem. 73 (1977) 313–325.

[14]   H. Burger, W. Doerfler: Intracellular forms of adenovirus DNA. III. Integration of the DNA of adenovirus type 2 into host DNA in productively infected cells. J. Virol. 13 (1974) 975–992.

[15]   T. B. Hasson, P. D. Soloway, D. Ornelles, W. Doerfler, T. Shenk: Adenovirus L1–52/55 kd proteins are required for assembly of virions. J. Virol. 63 (1989) 3612–3621.

[16]   U. Müller, M. P. Roberts, D. A. Engel, W. Doerfler, T. Shenk: Induction of transcription factor AP-1 by adenovirus E1A protein and cAMP. Genes Devel. 3 (1989) 1991–2002.

[17]   U. Wienhues, K. Hosokawa, A. Höveler, B. Siegmann, W. Doerfler: A novel method for the transcription and expression of reconstituted DNA-protein complexes in eukaryotic cells. DNA 6 (1987) 81–89.

*Walter Doerfler*

[18] A. J. van der Eb, H. T. M. Timmers, R. Offringa, A. Zantema, S. J. L. van den Heuvel, J. A. F. van Dam, J. L. Bos: Suppression of cellular gene activity in adenovirus-transformed cells. Curr. Top. Microbiol. Immunol. 144 (1989) 197–207.

[19] P. Whyte, K. J. Buchkovich, J. M. Horowitz, S. H. Friend, M. Raybuck, R. A. Weinberg, E. Harlow: Association between an oncogene and an anti-oncogene: the adenovirus E1A proteins bind to the retinoblastoma gene product. Nature 334 (1988) 124–129.

[20] I. Kuhlmann, S. Achten, R. Rudolph, W. Doerfler: Tumor induction by human adenovirus type 12 in hamsters: loss of the viral genome from adenovirus type 12-induced tumor cells is compatible with tumor formation. EMBO J. 1 (1982) 79–86.

[21] R. Gahlmann, W. Doerfler: Integration of viral DNA into the genome of the adenovirus type 2-transformed hamster cell line HE5 without loss or alteration of cellular nucleotides. Nucleic Acids Res. 11 (1983) 7347–7361.

[22] M. Schulz, W. Doerfler: Deletion of cellular DNA at site of viral DNA insertion in the adenovirus type 12-induced mouse tumor CBA-12-1-T. Nucleic Acids Res. 12 (1984) 4959–4976.

[23] J. Groneberg, D. Sutter, H. Soboll, W. Doerfler: Morphological revertants of adenovirus type 12-transformed hamster cells. J. Gen. Virol. 40 (1978) 635–645.

[24] R. Deuring, U. Winterhoff, F. Tamanoi, S. Stabel, W. Doerfler: Site of linkage between adenovirus type 12 and cell DNAs in hamster tumour line CLAC3. Nature 293 (1981) 81–84.

[25] R. Gahlmann, R. Leisten, L. Vardimon, W. Doerfler: Patch homologies and the integration of adenovirus DNA in mammalian cells. EMBO J. 1 (1982) 1101–1104.

[26] S. Stabel, W. Doerfler: Nucleotide sequence at the site of junction between adenovirus type 12 DNA and repetitive hamster cell DNA in transformed cell line CLAC1. Nucleic Acids Res. 10 (1982) 8007–8023.

[27] W. Doerfler, R. Gahlmann, S. Stabel, R. Deuring, U. Lichtenberg, M. Schulz, D. Eick, R. Leisten: On the mechanism of recombination between adenoviral and cellular DNAs: the structure of junction sites. Curr. Top. Microbiol. Immunol. 109 (1983) 193–228.

[28] U. Lichtenberg, C. Zock, W. Doerfler: Insertion of adenovirus type 12 DNA in the vicinity of an intracisternal A particle genome in Syrian hamster tumor cells. J. Virol. 61 (1987) 2719–2726.

[29] R. Jessberger, B. Weisshaar, S. Stabel, W. Doerfler: Arrangement and expression of integrated adenovirus type 12 DNA in the transformed hamster cell line HA12/7: amplification of Ad12 and c-myc DNAs and evidence for hybrid viral-cellular transcripts. Virus Res. 13 (1989) 113–128.

[30] R. Deuring, G. Klotz, W. Doerfler: An unusual symmetric recombinant between adenovirus type 12 DNA and human cell DNA. Proc. Natl. Acad. Sci. USA 78 (1981) 3142–3146.

[31] R. Deuring, W. Doerfler: Proof of recombination between viral and cellular genomes in KB cells productively infected by adenovirus type 12: structure of the junction site in a symmetric recombinant (SYREC). Gene 26 (1983) 283–289.

[32] W. Doerfler, H. Burger, J. Ortin, E. Fanning, D. T. Brown, M. Westphal, U. Winterhoff, B. Weiser, J. Schick: Integration of adenovirus DNA into the cellular genome. Cold Spring Harbor Symp. Quant. Biol. 39 (1974) 505–521.

262

[33] J. Schick, K. Baczko, E. Fanning, J. Groneberg, H. Burger, W. Doerfler: Intracellular forms of adenovirus DNA: Integrated form of adenovirus DNA appears early in productive infection. Proc. Natl. Acad. Sci. USA 73 (1976) 1043–1047.

[34] K. Baczko, R. Neumann, W. Doerfler: Intracellular forms of adenovirus DNA. VII. Excision of viral sequences from cellular DNA in adenovirus type 2-infected KB cells. Virology 85 (1978) 557–567.

[35] W. Doerfler, S. Stabel, H. Ibelgaufts, D. Sutter, R. Neumann, J. Groneberg, K. H. Scheidtmann, R. Deuring, U. Winterhoff: Selectivity in integration sites of adenoviral DNA. Cold Spring Harbor Symp. Quant. Biol. 44 (1979) 551–564.

[36] R. Neumann, W. Doerfler: Integration of adenovirus type 2 DNA at a limited number of cellular sites in productively infected cells. J. Virol. 37 (1981) 887–892.

[37] T. Rosahl, W. Doerfler: Predominant association of adenovirus type 12 DNA with human chromosome 1 early in productive infection. Virology 162 (1988) 494–497.

[38] R. Gahlmann, M. Schulz, W. Doerfler: Low molecular weight RNAs with homologies to cellular DNA at sites of adenovirus DNA insertion in hamster or mouse cells. EMBO J. 3 (1984) 3263–3269.

[39] M. Schulz, U. Freisem-Rabien, R. Jessberger, W. Doerfler: Transcriptional activities of mammalian genomes at sites of recombination with foreign DNA. J. Virol. 61 (1987) 344–353.

[40] D. Eick, S. Stabel, W. Doerfler: Revertants of adenovirus type 12-transformed hamster cell line T637 as tools in the analysis of integration patterns. J. Virol. 36 (1980) 41–49.

[41] I. Kuhlmann, W. Doerfler: Shifts in the extent and patterns of DNA methylation upon explantation and subcultivation of adenovirus type 12-induced hamster tumor cells. Virology 118 (1982) 169–180.

[42] I. Kuhlmann, W. Doerfler: Loss of viral genomes from hamster tumor cells and nonrandom alterations in patterns of methylation of integrated adenovirus type 12 DNA. J. Virol. 47 (1983) 631–636.

[42a] G. Orend, I. Kuhlmann, W. Doerfler: Spreading of DNA methylation across integrated foreign (adenovirus type 12) genomes in mammalian cells. J. Virol. 65 (1991) 4301–4308.

[43] J. Groneberg, Y. Chardonnet, W. Doerfler: Integrated viral sequences in adenovirus type 12-transformed hamster cells. Cell 10 (1977) 101–111.

[44] D. Sutter, M. Westphal, W. Doerfler: Patterns of integration of viral DNA sequences in the genomes of adenovirus type 12-transformed hamster cells. Cell 14 (1978) 569–585.

[45] S. Stabel, W. Doerfler, R. R. Friis: Integration sites of adenovirus type 12 DNA in transformed hamster cells and hamster tumor cells. J. Virol. 36 (1980) 22–40.

[46] S. Vogel, T. Rosahl, W. Doerfler: Chromosomal localization of integrated adenovirus DNA in productively infected and in transformed mammalian cells. Virology 152 (1986) 159–170.

[47] D. Eick, B. Kemper, W. Doerfler: Excision of amplified viral DNA at palindromic sequences from the adenovirus type 12-transformed hamster cell line T637. EMBO J. 2 (1983) 1981–1986.

[48] E. Fanning, W. Doerfler: Intracellular forms of adenovirus DNA. VI. Quantitation and characterization of the four size-classes of adenovirus type 2 DNA in human KB cells. Virology 81 (1977) 433–448.

[49] W. Doerfler: Integration of viral DNA into the host genome. Current Topics Microbiol. Immunol. 71 (1975) 1–78.

[50] W. Doerfler: Animal virus-host genome interactions. In: H. Fraenkel-Conrat, R. R. Wagner (eds.): Comprehensive Virology. Plenum Press, New York 1977, Vol. 10, pp. 279–399.

[51] W. Doerfler, A. Spies, R. Jessberger, U. Lichtenberg, C. Zock, T. Rosahl: Recombination of foreign (viral) DNA with the host genome. Studies in vivo and in a cell free system. In: R. Rott, W. Goebel (eds.): Molecular Basis of Viral and Microbial Pathogenesis. 38. Colloquium Mosbach 1987. Springer-Verlag, Berlin-Heidelberg 1987, pp. 60–72.

[52] R. Jessberger, D. Heuss, W. Doerfler: Recombination in hamster cell nuclear extracts between adenovirus type 12 DNA and two hamster preinsertion sequences. EMBO J. 8 (1989) 869–878.

[53] W. A. Strohl, H. C. Rouse, R. W. Schlesinger: Properties of cells derived from adenovirus-induced hamster tumors by long-term in vitro cultivation. II. Nature of the restricted response to type 2 adenovirus. Virology 28 (1966) 645–658.

[54] W. A. Strohl: The response of BHK21 cells to infection with type 12 adenovirus. I. Cell killing and T antigen synthesis as correlated viral genome functions. Virology 39 (1969) 642–652.

[55] E. Fanning, W. Doerfler: Intracellular forms of adenovirus DNA. V. Viral DNA sequences in hamster cells abortively infected and transformed with human adenovirus type 12. J. Virol. 20 (1976) 373–383.

[56] K. Raska, Jr., W. A. Strohl: The response of BHK21 cells to infection with type 12 adenovirus. VI. Synthesis of virus-specific RNA. Virology 47 (1972) 734–742.

[57] J. Ortin, W. Doerfler: Transcription of the genome of adenovirus type 12. I. Viral mRNA in abortively infected and transformed cells. J. Virol. 15 (1975) 27–35.

[58] J. Ortin, K.-H. Scheidtmann, R. Greenberg, M. Westphal, W. Doerfler: Transcription of the genome of adenovirus type 12. III. Maps of stable RNA from productively infected human cells and abortively infected and transformed hamster cells. J. Virol. 20 (1976) 355–372.

[59] T. Klimkait, W. Doerfler: Adenovirus types 2 and 5 functions elicit replication and late expression of adenovirus type 12 DNA in hamster cells. J. Virol. 55 (1985) 466–474.

[60] T. Klimkait, W. Doerfler: E1B functions of type C adenoviruses play a role in the complementation of blocked adenovirus type 12 DNA replication and late gene transcription in hamster cells. Virology 161 (1987) 109–120.

[61] H. Esche, R. Schilling, W. Doerfler: In vitro translation of adenovirus type 12-specific mRNA isolated from infected and transformed cells. J. Virol. 30 (1979) 21–31.

[62] H. Esche, B. Siegmann: Expression of early viral gene products in adenovirus type 12-infected and -transformed cells. J. Gen. Virol. 60 (1982) 99–113.

[63] H. Esche, M. Reuther, K. Schughart: Early and late proteins of adenovirus type 12: Translation mapping with RNA isolated from infected and transformed cells. Curr. Top. Microbiol. Immunol. 111 (1984) 91–106.

[64] R. Jüttermann, U. Weyer, W. Doerfler: On the defect of adenovirus type 12 replication in hamster cells: absence of transcription of viral VA and L1 RNAs. J. Virol. 63 (1989) 3535–3540.

[65] L. T. Chow, T. R. Broker, J. B. Lewis: Complex splicing patterns of RNAs from the early regions of adenovirus-2. J. Mol. Biol. 134 (1979) 265–303.

[66] A. R. Shaw, E. B. Ziff: Transcripts from the adenovirus-2 major late promoter yield a single early family of 3′ coterminal mRNAs and five late families. Cell 22 (1980) 905–916.

[67] G. Akusjärvi, H. Persson: Controls of RNA splicing and termination in the major late adenovirus transcription unit. Nature 292 (1981) 420–426.

[68] U. Weyer, W. Doerfler: Species dependence of the major late promoter in adenovirus type 12 DNA. EMBO J. 4 (1985) 3015–3019.

[69] C. Zock, W. Doerfler: A mitigator sequence in the downstream region of the major late promoter of adenovirus type 12 DNA. EMBO J. 9 (1990) 1615–1623.

[70] L. Visser, A. C. M. B. Reemst, A. D. M. van Mansfeld, T. H. Rozijn: Nucleotide sequence analysis of the linked left and right hand terminal regions of adenovirus type 5 DNA present in the transformed rat cell line 5RK20. Nucleic Acids Res. 10 (1982) 2189–2198.

[71] B. Weisshaar, K.-D. Langner, R. Jüttermann, U. Müller, C. Zock, T. Klimkait, W. Doerfler: Reactivation of the methylation-inactivated late E2A promoter of adenovirus type 2 by E1A (13S) functions. J. Mol. Biol. 202 (1988) 255–270.

[72] D. Sutter, W. Doerfler: Methylation of integrated viral DNA sequences in hamster cells transformed by adenovirus 12. Cold Spring Harbor Symp. Quant. Biol. 44 (1979) 565–568.

[73] D. Sutter, W. Doerfler: Methylation of integrated adenovirus type 12 DNA sequences in transformed cells is inversely correlated with viral gene expression. Proc. Natl. Acad. Sci. USA 77 (1980) 253–256.

[74] L. Vardimon, R. Neumann, I. Kuhlmann, D. Sutter, W. Doerfler: DNA methylation and viral gene expression in adenovirus-transformed and -infected cells. Nucleic Acids Res. 8 (1980) 2461–2473.

[75] I. Kruczek, W. Doerfler: The unmethylated state of the promoter/leader and 5′-regions of integrated adenovirus genes correlates with gene expression. EMBO J. 1 (1982) 409–414.

[76] W. Doerfler: DNA methylation and gene activity. Ann. Rev. Biochem. 52 (1983) 93–124.

[77] W. Doerfler: DNA methylation – a regulatory signal in eukaryotic gene expression. J. Gen. Virol. 57 (1981) 1–20.

[78] W. Doerfler: DNA methylation: role in viral transformation and persistence. Adv. Viral Oncol. 4 (1984) 217–247.

[79] W. Doerfler: Complexities in gene regulation by promoter methylation. In: F. Eckstein, D. M. Lilley (eds.): Nucleic Acids and Molecular Biology. Springer-Verlag, Berlin-Heidelberg-New York-London-Paris-Tokyo 1989, Vol. 3, pp. 92–119.

[80] W. Doerfler: The significance of DNA methylation patterns: Promoter inhibition by sequence-specific methylation is one functional consequence. Phil. Transact. Royal Society, London, B 326 (1990) 253–265.

[81] W. Doerfler: DNA methylation and gene activity. In: Encyclopedia of Human Biology, Academic Press 1991, pp. 151–162.

[82] I. Kruczek, W. Doerfler: Expression of the chloramphenicol acetyltransferase gene in mammalian cells under the control of adenovirus type 12 promoters: effect of promoter methylation on gene expression. Proc. Natl. Acad. Sci. USA 80 (1983) 7586–7590.

[83] D. Knebel, W. Doerfler: $N^6$-Methyldeoxyadenosine residues at specific sites decrease the activity of the E1A promoter of adenovirus type 12 DNA. J. Mol. Biol. 189 (1986) 371–375.

[84] L. Vardimon, A. Kressmann, H. Cedar, M. Maechler, W. Doerfler: Expression of a cloned adenovirus gene is inhibited by in vitro methylation. Proc. Natl. Acad. Sci. USA 79 (1982) 1073–1077.

[85] K.-D. Langner, L. Vardimon, D. Renz, W. Doerfler: DNA methylation of three 5' C-C-G-G 3' sites in the promoter and 5' region inactivates the E2a gene of adenovirus type 2. Proc. Natl. Acad. Sci. USA 81 (1984) 2950–2954.

[86] K.-D. Langner, U. Weyer, W. Doerfler: Trans effect of the E1 region of adenoviruses on the expression of a prokaryotic gene in mammalian cells: resistance to 5'-CCGG-3' methylation. Proc. Natl. Acad. Sci. USA 83 (1986) 1598–1602.

[87] G. L. Cantoni, A. Razin (eds.): Biochemistry and Biology of DNA Methylation. Alan R. Liss, New York 1985.

[88] D. Knebel, H. Lübbert, W. Doerfler: The promoter of the late p10 gene in the insect nuclear polyhedrosis virus Autographa californica: activation by viral gene products and sensitivity to DNA methylation. EMBO J. 4 (1985) 1301–1306.

[89] U. Müller, W. Doerfler: Fixation of unmethylated or 5'-CCGG-3' methylated foreign DNA in the genome of hamster cells: gene expression and stability of methylation patterns. J. Virol. 61 (1987) 3710–3720.

[90] M. Toth, U. Müller, W. Doerfler: The establishment of de novo DNA methylation patterns: Transcription factor binding and deoxycytidine methylation at CpG and non-CpG sequences in an integrated adenovirus promoter. J. Mol. Biol. 214 (1990) 673–683.

[91] P. Dobrzanski, A. Hoeveler, W. Doerfler: Inactivation by sequence-specific methylations of adenovirus promoters in a cell-free transcription system. J. Virol. 62 (1988) 3941–3946.

[92] P. A. Jones, S. M. Taylor: Cellular differentiation, cytidine analogs and DNA methylation. Cell 20 (1980) 85–93.

[93] M. Groudine, R. Eisenman, H. Weintraub: Chromatin structure of endogenous retroviral genes and activation by an inhibitor of DNA methylation. Nature 292 (1981) 311–317.

[94] B. Knust, U. Brüggemann, W. Doerfler: Reactivation of a methylation-silenced gene in adenovirus-transformed cells by 5-azacytidine or by E1A transactivation. J. Virol. 63 (1989) 3519–3524.

[95] D. Knebel-Mörsdorf, S. Achten, K.-D. Langner, R. Rüger, B. Fleckenstein, W. Doerfler: Reactivation of the methylation-inhibited late E2A promoter of adenovirus type 2 DNA by a strong enhancer of human cytomegalovirus. Virology 166 (1988) 166–174.

[96] M. Toth, U. Lichtenberg, W. Doerfler: Genomic sequencing reveals a 5-methyldeoxycytosine-free domain in active promoters and the spreading of preimposed methylation patterns. Proc. Natl. Acad. Sci. USA 86 (1989) 3728–3732.

[97] S. M. Gartler, A. D. Riggs: Mammalian X-chromosome inactivation. Ann. Rev. Genet. 17 (1983) 155–190.

[98] R. Hermann, A. Hoeveler, W. Doerfler: Sequence-specific methylation in a downstream region of the late E2A promoter of adenovirus type 2 DNA interferes with protein binding. J. Mol. Biol. 210 (1989) 411–415.

[99] G. E. Smith, M. D. Summers, M. J. Fraser: Production of human beta interferon in insect cells infected with a baculovirus expression vector. Mol. Cell. Biol. 3 (1983) 2156–2165.

[100] E. B. Carstens, S. T. Tjia, W. Doerfler: Infection of Spodoptera frugiperda cells with Autographa californica nuclear polyhedrosis virus. I. Synthesis of intracellular proteins after virus infection. Virology 99 (1979) 386–398.

[101] S. T. Tjia, E. B. Carstens, W. Doerfler: Infection of Spodoptera frugiperda cells with Autographa californica nuclear polyhedrosis virus. II. The viral DNA and the kinetics of its replication. Virology 99 (1979) 399–409.

[102] H. Lübbert, W. Doerfler: Mapping of early and late transcripts encoded by the Autographa californica nuclear polyhedrosis virus genome: Is viral RNA spliced? J. Virol. 50 (1984) 497–506.

[103] H. Esche, H. Lübbert, B. Siegmann, W. Doerfler: The translational map of the Autographa californica nuclear polyhedrosis virus (AcNPV) genome. EMBO J. 1 (1982) 1629–1633.

[104] H. Lübbert, W. Doerfler: Transcription of overlapping sets of RNAs from the genome of Autographa californica nuclear polyhedrosis virus: a novel method for mapping RNAs. J. Virol. 52 (1984) 255–265.

[105] C. Oellig, B. Happ, T. Müller, W. Doerfler: Overlapping sets of viral RNAs reflect the array of polypeptides in the EcoRI J and N fragments (map positions 81.2 to 85.0) of the Autographa californica nuclear polyhedrosis virus genome. J. Virol. 61 (1987) 3048–3057. C. Oellig, B. Happ, T. Müller, W. Doerfler: Author's correction. J. Virol. 63 (1989) 1494.

[106] V. A. Luckow, M. D. Summers: Trends in the development of baculovirus expression vectors. BioTechnology 6 (1988) 47–55.

[107] S. T. Tjia, G. Meyer zu Altenschildesche, W. Doerfler: Autographa californica nuclear polyhedrosis virus (AcNPV) DNA does not persist in mass cultures of mammalian cells. Virology 125 (1983) 107–117.

[108] K. Kuroda, C. Hauser, R. Rott, H.-D. Klenk, W. Doerfler: Expression of the influenza virus haemagglutinin in insect cells by a baculovirus vector. EMBO J. 5 (1986) 1359–1365.

[109] K. Kuroda, A. Gröner, K. Frese, C. Hauser, R. Rott, W. Doerfler, H.-D. Klenk: Expression of the influenza virus hemagglutinin in insect larvae by a baculovirus vector. J. Virol. 63 (1989) 1677–1685.

[110] C. Hauser, H. Fußwinkel, J. Li, C. Oellig, R. Kunze, M. Müller-Neumann, M. Heinlein, P. Starlinger, W. Doerfler: Overproduction of the protein encoded by the maize transposable element Ac in insect cells by a baculovirus vector. Mol. Gen. Genet. 214 (1988) 373–378.

[111] K. Kuroda, H. Geyer, R. Geyer, W. Doerfler, H.-D. Klenk: The oligosaccharides of influenza virus hemagglutinin expressed in insect cells by a baculovirus vector. Virology 174 (1990) 418–429.

*Walter Doerfler*

# Survey of Doctoral Students and Postdoctoral Scientists

(* Diploma in W. D.'s laboratory)

## Doctoral Students

Byron T. Burlingham (1967–1970, Ph. D. Rockefeller University): Characteristics of an endonuclease associated with adenovirus type 2 and 12. Professor of Microbiology, Medical School, University of North Carolina, USA.

Harold Burger (1970–1974, Ph. D. Rockefeller University): The integration of adenovirus DNA into the DNA of productively infected human cells. Albert Einstein College of Medicine; Postdoctoral Fellow Stanford University, USA; Assistant Professor, SUNY, Med. School., Stony Brook, N. Y., USA.

Ellen Fanning* (1972–1976): Viral DNA sequences in adenovirus-infected and -transformed cells. Universität Konstanz, FRG; Professor of Biochemistry, Universität München, FRG.

Ulrike Reif (1972–1977): Isolierung und Charakterisierung von Endonukleasen aus uninfizierten und Adenovirus Typ 2 infizierten KB Zellen. Teaching in Bakersfield, Calif., USA.

Jürgen Groneberg* (1972–1979): Isolierung und Charakterisierung von morphologischen Revertanten Adenovirus Typ 12-transformierter Hamsterzellen. Fa. Hoechst, Frankfurt, FRG.

Sian T. Tjia* (1972–1979): Wechselwirkungen von *Autographa californica* Kernpolyeder-Viren (AcNPV) mit Insekten- und Säugerzellen. Akademischer Oberrat, Universität Köln, FRG.

Karl-Heinz Scheidtmann* (1973–1978): Transcription des Adenovirus-Typ-12 Genoms: Charakterisierung und Kartierung der virusspezifischen RNA aus produktiv oder abortiv infizierten sowie aus transformierten Zellen. Priv. Doz., Heisenberg-Stipendiat, Universität Freiburg, FRG; Professor of Genetics, Universität Bonn, FRG.

Jochen Schick* (1973–1978): Charakterisierung der integrierten Form der Adenovirus DNS in produktiv infizierten Zellen. Amersham-Buchler, Kopenhagen, Frankfurt; Sanorell Pharma, Baiersbronn, FRG.

Horst Ibelgaufts (1975–1979): Charakterisierung Adenovirus Typ 12-induzierter neuroepithelialer Tumoren aus BDIX Ratten. Assistent, Universität München, FRG.

Reiner Neumann* (1975–1981): Interaktion zwischen viraler und zellulärer DNA in Adenovirus infizierten und transformierten Säugerzellen. Institut für Virologie, Universität zu Köln, FRG; Bayer AG.

Lily Vardimon (1976–1982): The interaction of adenovirus types 2 and 12 with *Muntiacus muntjak* cells. MIT, Cambridge, USA; Assistant Professor, University of Chicago, USA; Professor of Microbiology, Tel Aviv University, Israel.

Diane Sutter (1976–1979): Patterns of integration of viral DNA sequences in the genomes of adenovirus type 12-transformed hamster cells and methylation of integrated viral DNA. UC Medical School, Denver, Colorado, USA.

Silvia Stabel* (1976–1982): Integrierte Adenovirus DNA in transformierten Hamsterzellen: Insertionsmuster und Nukleotidsequenz an der Insertionsstelle. EMBL, Heidelberg, FRG; Ludwig Institute, London, UK.; Group Leader, Max-Delbrück-Laboratorium, Köln, FRG.

Dirk Eick* (1978–1983): Integration und Excision viraler DNA in Adenovirus Typ 12 transformierten Hamsterzellen und Untersuchungen am Methylierungsmuster der viralen DNA. Priv. Doz. Universität Freiburg, FRG; Group leader GSF München, FRG.

Renate Deuring* (1978–1983): Rekombination zwischen viraler und zellulärer DNA: Im Chromosom transformierter Zellen integriert und im Virion verpackt. Postdoctoral Fellow, MIT, Cambridge, USA; University of Colorado, Boulder, USA; UC Santa Cruz, USA.

Ingrid Kuhlmann (1979–1983): Tumorinduktion durch Adenovirus Typ 12 in neugeborenen Hamstern. Associate Director, Animal Research, University of Konstanz, FRG.

Reinhold Gahlmann* (1979–1984): Charakterisierung der Integrationsstelle und des Integrationsmusters viraler DNA im Genom der Adenovirus Typ 2-transformierten Hamsterzellinie HE5. Postdoctoral Fellow, Stanford University; Assistant Professor, University of Southern California, Los Angeles, USA; Bayer AG, Wuppertal.

Hermann Lübbert* (1979–1984): Das *Autographa californica* Kernpolyedervirus als Modellsystem für die Untersuchung molekularbiologischer Mechanismen: Untersuchungen an Struktur und Organisation des viralen Genoms. Postdoctoral Fellow, California Institute of Technology, Pasadena, USA; Group leader Fa. Sandoz, Basel, Switzerland.

Sebastian Vogel* (1979–1985): Chromosomale Lokalisierung integrierter Adenovirus DNA in transformierten und produktiv infizierten Zellen. Science journalism.

Sabine Achten* (1979–1986): Charakterisierung einer hoch repetitiven DNA-Sequenz der Ratte mit langen offenen Leserastern. "Assistentin", Universtät zu Köln, FRG; Boehringer-Ingelheim Fonds, Stuttgart, FRG.

Manfred Schulz* (1980–1986): Integration von Adenovirus DNA in das Säugergenom an transkriptionsaktiven Stellen. Postdoctoral Fellow, University of Zürich, Switzerland, Sandoz, Basel, Switzerland.

Birgitt Knust* (1981–1988): Reaktivierung eines genomisch fixierten methylierten viralen Promoters in Hamsterzellen: der späte E2A-Promotor in der Adenovirus Typ 2 transformierten Hamsterzellinie HE3. Postdoctoral Fellow, Carlsberg Laboratories, Kopenhagen, Denmark.

Klaus-Dieter Langner* (1981–1986): Inaktivierung des Ad2 E2A Promotors durch sequenzspezifische DNA Methylierung: Einfluß von Hemimethylierung, von Transaktivierung und von viralen "enhancer" Elementen. Behring-Werke, Marburg, FRG.

Ulrike Weyer* (1981–1986): Transaktivierung von Promotoren durch species-spezifische zelluläre Faktoren und durch Adenovirus Genprodukte. Postdoctoral Fellow, Oxford University, Oxford, UK; Bender & Co., Vienna, Austria.

Thomas Klimkait* (1981–1986): Adenovirus Typ 2- und Typ 5-Funktionen komplementieren blockierte DNA-Replikation und -Expression von Adenovirus Typ 12 in Hamsterzellen. Postdoctoral Fellow, National Institutes of Health, Bethesda, Maryland, USA; CIBA-GEIGY, Basel, Switzerland.

Ursula Lichtenberg* (1981–1988): Instabilität fremder DNA in Adenovirus Typ 12-induzierten Hamster Tumorzellen: Struktur einer viralen Integrationsstelle. Postdoctoral Fellow, University of California, San Francisco, USA.

Cornelia Oellig* (1982–1989): Besonderheit des Transkriptionsmodus des *Autographa californica* Kernpolyedervirus. Postdoctoral Fellow, Karolinska Institutet, Stockholm, Sweden.

Pawel Dobrzanski (1982–1988): Inaktivierung durch Sequenz-spezifische Methylierung von Adenovirus Promotoren in einem zellfreien Transkriptionssystem. Postdoctoral Fellow, EMBL, Heidelberg, FRG; Squibb Research Laboratories, Princeton, USA.

Dagmar Knebel-Mörsdorf (1983–1986): Virale Promotoren: Inaktivierung durch sequenz-spezifische Methylierungen und Transaktivierung in heterologen Wirtssystemen. Postdoctoral Fellow, Universität zu Köln, FRG; Group Leader, Sonderforschungsbereich 274, Universität zu Köln, FRG.

Rolf Jessberger* (1983–1989): Untersuchungen zur Rekombination zwischen viraler und zellulärer DNA: Studien *in vivo* und in einem zellfreien System. Postdoctoral Fellow, Department of Biochemistry, Stanford University, Stanford, USA.

Arnd Höveler* (1983–1988): Inaktivierung des späten E2A Promotors von Adenovirus Typ 2 DNA durch sequenz-spezifische Methylierung: DNA-Protein Bindung und Promotorinaktivierung in einem zellfreien Transkriptionssystem. Postdoctoral Fellow, INSERM, Marseille, France; Professor, Univ. of Besançon, France.

Ulla Wienhues* (1983–1988): Aufnahme von DNA-Protein-Komplexen in Säugerzellen: Entwicklung einer neuen, effizienten Transfektionsmethode. Postdoctoral Fellow, Institute of Physiological Chemistry, Universität München, FRG; Boehringer Tutzing, FRG.

Brigitte Happ* (1984–1990): In vitro Translation und Charakterisierung im *Autographa californica* Kernpolyedervirus Genom encodierter Proteine. Postdoctoral Fellow, Friedrich Miescher Institut, Basel, Switzerland.

Bernd Weisshaar (1985–1988): Inhibition eines viralen Promotors durch Sequenzenspezifische Methylierung: Aufhebung des Transkriptionsblockes durch E1A-Transaktivierung. Postdoctoral Fellow, Gruppenleiter, Max-Planck-Institut für Züchtungsforschung, Köln, FRG.

Jian Li (1988–1990): The *Autographa californica* nuclear polyhedrosis virus (AcNPV) expression vector system: Studies on the transposable element Ac from *Zea mays,* the 81.2 to 85.0 map unit segment of AcNPV DNA and on the oncogene jun of avian sarcoma virus 17. Postdoctoral Fellow, University of Southern California, Los Angeles, USA.

Jochen Schellner* (1982–1990): Untersuchungen zur "heterologen" Rekombination in einem zellfreien System aus Säugerzellkernen. Asta Werke, Hamburg, FRG.

Charlotte Hauser* (1984–1990): Studien zur allgemeinen homologen Rekombination in eukaryontischen Zellen. A) Integration durch homologe Rekombination des Ac Elementes von *Zea mays* L. in einem Baculovirus Vektor und seine Expression in Insektenzellen. B) Identifikation und Charakterisierung von Rekombinationsproteinen in höheren eukaryontischen Zellen. (Dissertation partly at the Massachusetts Institute of Technology) Institut Pasteur, Paris, France.

Albrecht Dehmel (1984–1990): Untersuchungen zum Methylierungsstatus des TNF$\alpha$-Gens sowie benachbarter Systeme. Training as Patent Lawyer, München, FRG.

Thomas Rosahl* (1984–1991): Regulation der viralen und zellulären Genexpression in Adenovirus-infizierten und -transformierten Zellen. Postdoctoral fellow, Southwestern Medical Center, Dallas, Texas, USA.

Ulrich Müller* (1985–1991): Induction of transcription factor AP-1 by adenovirus E1A protein and cAMP. (Dissertation at Princeton University) Postdoctoral fellow, University of California, Berkeley, California, USA.

Clara Maria Lettmann (1987–1991): Untersuchung zur Methylierung des späten E2A-Promotors von Adenovirus Typ 2 in transgenen Mäusen. Fa. Braun, Melsungen, FRG.

270

Ruth Jüttermann (1981–1991): Die Inhibition der L1- und VAI-Gene von Adenoviren in nicht-permissiven Zellen und des VAI-Gens nach Sequenz-spezifischer Methylierung. Postdoctoral Fellow, The Whitehead Institute, Massachusetts, USA.

Ralf Hermann* (1987–1992): Zum Mechanismus der Promotorinaktivierung durch sequenzspezifische DNA-Methylierung: Untersuchungen am späten E2A-Promotor von Adenovirus Typ 2. Diagen, Düsseldorf, FRG.

## Postdoctoral Scientists

Monika Hirsch-Kauffmann (1969–1971), Professor of Genetics, Universität Innsbruck, Austria.

Brigitte Rosenwirth (1972–1974), Sandoz Wien, Austria.

Juan Ortin* (1972–1975), Research Professor, Universidad Autonomà Madrid, Spain.

Knut Baczko (1975–1978), Priv. Doz., Universität Würzburg, FRG.

Helmut Esche (1976–1979), Professor of Molecular Biology, Universität Essen, FRG.

Eric B. Carstens* (1977–1979), Professor of Microbiology, Queen's University, Kingston, Ontario, Canada.

Fuyuhiko Tamanoi (1980), Cold Spring Harbor; Associate Professor, University of Chicago, USA.

Inge Kruczek (1981–1983), Akademischer Rat, Universität München, FRG; Zentrale Kommission für die Biologische Sicherheit, Berlin.

Basil Arif* (1982–1984), Staff Member, Forest Pest Management Institute, Sault Ste. Marie, Ontario, Canada.

David Zarling* (1980–1981), USA.

Pei Mei-yun* (1985–1986), Academia Sinica, Beijing, China (deceased).

Thomas Müller (1984–1986), Gesellschaft für Biologische Forschung, Köln, FRG.

Ewa Bartnik* (1985–1987), Dozentin, University of Warszawa, Poland.

Sabine Achten (since 1986), "Assistentin", Universität Köln, FRG, Boehringer Ingelheim Fonds, Stuttgart, FRG.

Dagmar Knebel (1986–1988), Group leader Sonderforschungsbereich 274, Institut für Genetik, Universität zu Köln, FRG.

Miklos Toth* (Ungarn, 1988–1990), Department of Molecular Biology, Howard Hughes Medical Institute, Princeton University, Princeton, USA.

Xiong Guangming* (China, 1988–1990); Universität Gießen, FRG.

Stefan Kochanek (1988–1992).

David Carson* (1990–1992). University of Texas, San Antonio, TX, USA.

---

*Fellows of the Alexander von Humboldt Foundation

## Habilitations

Dr. Ernst-Ludwig Winnacker (1976), Professor of Biochemistry, Director Genzentrum, Universität München, FRG.

Dr. Helmut Esche (1982), Professor of Molecular Biology, Universität Essen, FRG.

Dr. Anna Starzinski-Powitz (1989), Group leader, Priv. Doz., Institut für Genetik, Universität zu Köln, FRG; Professor of Human Genetics, University of Frankfurt, FRG.

# 6.2 Adenovirus DNA: Structure and Function of a Novel Replicon

Ernst-L. Winnacker*

## 6.2.1 Introduction

Viral nucleic acids have always served as useful model systems for the study of macromolecular biosynthesis, e. g. transcription, recombination and DNA replications. At the beginning of my tenure in the laboratory of Prof. W. Doerfler (1972–1976), the adenovirus genome appeared as a valuable choice for the study of DNA replication. It was available in large amounts and it was known to undergo an efficient and well-defined replication cycle in infected human cell lines. On the way towards the development of an *in vitro* system for adenovirus DNA replication, it was paramount to develop restriction maps of the adenovirus genome and to identify the *in vivo* mechanism of replication, including the location of origins and termini of DNA replication. With the approach of developing a set of conditional-lethal mutants temperature-sensitive for DNA synthesis, we hoped to distinguish cellular from viral factors required for this process.

## 6.2.2 Viral and Host Functions in the Lytic Infection Cycle

Adenovirus DNA synthesis occurs semiconservatively in the nucleus of the infected host cell. Its actual location within these confines has been difficult to establish.

The "early" part of the lytic infection cycle, which lasts until the onset of viral DNA replication, is characterized by the synthesis and expression of

---

* Institut für Genetik der Universität zu Köln
  (at present: Institut für Biochemie der Universität München)

"early" mRNA, which maps at four different regions of the viral genome. At least three "early" gene functions are defined by temperature-sensitive mutants deficient in viral DNA replication at the restrictive temperature.

One of these lesions occurs in a protein which binds preferentially to single-stranded DNA. This nonstructural protein of molecular weight 72,000 (Ad2 and Ad5) or 60,000 daltons (Ad12) has been shown to be a product of *in vitro* translation of purified mRNA derived from the Eco R1-B fragment of the viral DNA, and thus maps between 0.59 and 0.71 map units. Temperature shift experiments with mutant H5ts125 indicate a direct involvement of this 72,000 dalton protein in the initiation of viral DNA synthesis. In addition, antibodies raised against purified 72,000 dalton protein preparations have been shown to partially inhibit viral DNA synthesis isolated nuclei from Ad5-infected KB cells.

Since such an *in vitro* system is believed to be deficient in initiation of new rounds of viral DNA replication and only able to complete preexisting replicative intermediates, the DNA binding protein thus appears to be required both for initiation and elongation of growing DNA chains. Its mechanism of action in these different processes remains unknown. One might speculate, however, that it serves both to protect the large amounts of single-stranded viral DNA from nuclease digestion and to maintain it in a favorable conformation – for example, the panhandled structures (Fig. 6.2.1) – for complementary strand synthesis.

A second "early" lesion is not as well characterized. Again, the respective ts-mutants appear to be defective in initiation of viral DNA replication, although the respective gene products have not been identified. Under nonpermissive conditions, the ability of one of these (H5ts36) to transform rat embryo cells is greatly reduced. It is striking that this lesion maps within the leftward 30% of the Ad5 genome. This region of the viral DNA has been shown to be both necessary and sufficient for cell transformation and is detected as such in various Ad5-transformed cell lines.

Sera from hamsters bearing group C adenovirus tumors initiated by injection of the Ad5-transformed cell line 14b, which contains only the left-hand 40% of the viral genome, precipitate a protein of molecular weight 58,000 daltons from lytically infected monkey or human cells. In addition, it is known from *in vitro* translation studies that this early region at the left-hand molecular end codes for at least two distinct polypeptides of molecular weight 15,000 and 44,000–50,000 daltons. The relationship of these latter two proteins to the H5ts36 lesion, to the 58,000 dalton group C adenovirus tumor antigen, and to the 55,000 dalton protein associated with the molecular ends of the viral DNA remains unknown, although an identity of the latter two proteins has been implicated.

On the basis of interserotypic complementation experiments with mutants H5ts125 and H5ts36, mutant H2ts206 exemplifies a third "early" complementation group [1]. Temperature shift experiments indicate that

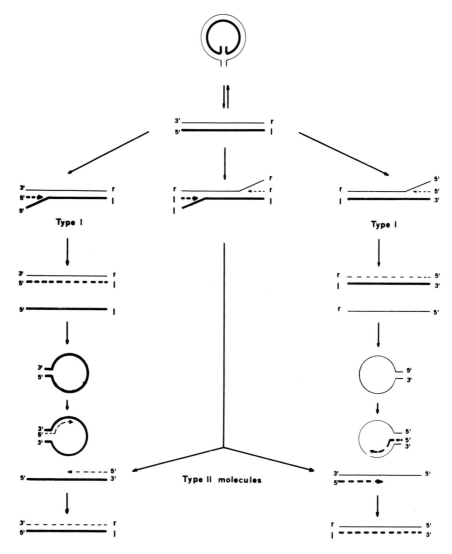

**Fig. 6.2.1:** Replication patterns of adenovirus DNA.

this lesion affects a function which is expressed within the first two hours after infection. Conceivably, it is not directly involved in viral DNA replication.

The three classes of Ad12 temperature-sensitive mutants are all defective in initiation of viral DNA replication. Their relationship to the Ad2/Ad5 mutants has not been determined. Since at least six distinct polypeptides have been identified as products of *in vitro* translation from early regions of the viral genome, genetic analysis of the "early" gene functions remains far from complete.

The observation that only very few viral gene functions appear to be associated with viral DNA replication indicates that this complex process depends largely upon the DNA synthesizing machinery of the host cell. At least three cellular proteins have been identified in the meantime to be required for initiation of viral DNA replication, Nuclear Factor I (NFI) and Nuclear Factor III (NFIII) [2].

## 6.2.3 Initiation of Adenovirus DNA Replication

The fundamental characteristics of the replication pattern of adenovirus DNA have been determined by electron microscopy, by an analysis of the nature of the single-strands DNA, and by elucidation of the temporal order of synthesis of specific regions of the viral genome.

Two basic classes of replicating DNA molecules have been recognized by electron microscopy: first, linear duplexes with one or two single-stranded branches (type I molecules; Fig. 6.2.1), and second, unbranched full-length molecules with a single- and double-stranded region (type II molecules; Figure 5.2.1). The double-stranded portions of both type I and type II molecules were shown by partial denaturation mapping to arise with equal frequency from both molecular ends of the viral genome. Only a small fraction of molecules consisted of structures containing features of both type I and II molecules – for example, linear structures with a variably located transition point between single- and double-stranded DNA, as well as with a single-stranded branch protruding from the double-stranded portion of the molecule. Together, these three forms account for 85–90 % of the total pool of replicating molecules.

These observations fit the following two-step model for adenovirus DNA synthesis (Fig. 6.2.1). In a first step, replication starts at or near either molecular end of the viral DNA by synthesis of a daughter strand in the 5′ to 3′ direction with concomitant displacement of the parental strand of the same polarity. In the case of initiation at the molecular right-hand end, the displaced strand is the parental r strand; in the case of initiation at the molecular left-hand end, it is the viral 1 strand. In contrast to a typical Cairns-type replication fork with two growing daughter strands, only one parental strand is replicated, and thus only one daughter molecule is produced at any given displacement fork. In this particular aspect, the proposed model is reminiscent of the similar asynchronous situation in mitochondrial DNA synthesis.

275

In the second step, a second daughter molecule is obtained by complementary strand synthesis from the 3′ end of the displaced parental strand. The "panhandled" structures proposed for the displaced single strands and formed by intramolecular base pairing of the inverted terminal sequences are hypothetical and have not been recognized as replicative intermediates. In principle, the duplex portions of "panhandled" single strands are identical in structure whether formed from r or 1 strands. They are thus both indistinguishable from each other and from the molecular right- or left-hand ends of duplex DNA for the enzymatic machinery responsible for recognition of the origins of DNA replication.

As pointed out by Lechner and Kelly [3], this aspect of the model emphasizes the structural and functional equivalence of all initiation events for displacement, as well as complementary strand synthesis from either of the molecular ends. In 1973, this latter aspect – the question of whether one or both molecular ends could serve as the origin for displacement synthesis – has been highly controversial. In previous studies, this problem was investigated using replicative intermediates, which had been either synthesized in a nuclear system [4] or obtained from virus-infected cells synchronized for viral DNA replication through release from a hydroxyurea block [5]. The results suggested that initiation of displacement synthesis occurs exclusively at the right-hand end of the viral DNA. A striking prediction of this model was the presence in infected cells of single-stranded DNA derived exclusively from r strand sequences. Although this had indeed been demonstrated by filter hybridization techniques, these results were challenged convincingly in both the Ad2 and the Ad5 system.

The present model (Fig. 6.2.1) is supported by a functional analysis on the location of origins and termini of adenovirus DNA replication through pulse label experiments. In these studies, termini of viral DNA replication were determined from an analysis of the distribution of newly incorporated radioactivity along the genome of mature viral DNA molecules, which were pulse-labeled for periods shorter than the time necessary for one round of viral DNA replication. The observation that such molecules exhibit two gradients for radioactivity directed from the center of the molecule toward both molecular ends is certainly consistent with the existence of two termini of replication at the respective molecular ends. The resolution of these experiments, which involved terminal fragments as small as 0.8 % of the viral genome (260 bp), places the termini within the 103 bp terminal repetition, if not toward the molecular ends themselves.

While this analysis identified termini of DNA replication, the results were still consistent with either an internal or a terminal location of the initiation sites. These mutually exclusive possibilities can be distinguished by an analysis of the strand specificity of pulse label in terminal restriction enzyme fragments. While internal initiation followed by the movement of two replication forks toward either molecular end would be expected to result in the accumulation of pulse label in both complementary strands of terminal restriction enzyme fragments, initiation at the molecular ends themselves and synthesis of daughter strands along the entire length of the parental strands would lead to an asymmetric unimodal distribution of label in pulse-labeled DNA. Actual analysis of the strand distribution of label in terminal restriction enzyme fragments by DNA/DNA hybridization or gel electrophoresis established the expected asymmetric distribution with >95% of the newly incorporated radioactivity in the strands with the respective 3′ molecular termini.

These studies were extended to include not only pulse-labeled mature viral DNA, but also replicating molecules. Thus it was possible by appropriate temperature-shift experiments with appropriate mutants to synchronize viral DNA replication and to demonstrate that label accumulates preferentially in terminal restriction enzyme fragments from both molecular ends of replicating viral DNA. In addition, origins of DNA replication were isolated directly from replicating DNA by a technique which involves sequential digestions of replicating DNA with restriction enzymes of different specificities and selection of fork-containing terminal fragments by chromatography on BND-cellulose.

## 6.2.4 Summary

Today, 15 years after completion of our work in Köln, the adenovirus system is still of considerable interest. *In vitro* systems for adenovirus DNA replication have been developed and used to characterize the various factor involved in adenovirus DNA replication. Two of the cellular factors, NFI and NFIII, which are sequestered by the infecting and replicating viral genome, have been characterized in considerable detail as to their DNA binding site and their role as transcriptional activators. The replication system can now be reconstituted completely from purified compounds and thus is apt to be good for many more surprises.

## 6.2.5  References

[1]  P. Kathmann, J. Schick, E. L. Winnacker and W. Doerfler: Isolation and characterization of temperature-sensitive mutants of adenovirus type 2. J. Virol. 19
(1976) 43–53.

[2]  M. Meisterernst, L. Rogge, R. Foeckler, M. Karaghiosoff and E. L. Winnacker:
Structural and functional organisation of a porcine gene coding for nuclear factor
I. Biochemistry 28 (1989) 8191–8200.

[3]  R. L. Lechner and T. J. Kelly Jr.: The structure of replicating adenovirus 2 DNA
molecules. Cell 12 (1977) 1007–1020.

[4]  J. S. Sussenbach, D. J. Ellens and H. J. Jansz: Studies on the mechanism of replication of adenovirus DNA. II. The nature of single-stranded DNA in replicative
intermediates. J. Virol. 12 (1973) 1131–1138.

[5]  D. J. Ellens, J. S. Sussenbach and H. S. Jansz: Studies on the mechanisms of replication of adenovirus DNA. III. Electron microscopy of replicating DNA. Virology 61 (1974) 427–442.

# 6.3 Malignant Transformation by Adenoviruses: Early Viral Gene Expression and Host Cell Response

Helmut Esche*

## 6.3.1 Introduction

The long range goals of our work are to achieve an understanding of the mechanism of virus-induced malignant cell transformation on a genetic and molecular level. The system we use to study cell transformation consists of adenoviruses that propagate in human cells permissive for virus growth (productive infection) but are unable to replicate in nonpermissive rodent cells. Some of these abortively infected rodent cells become transformed to malignancy by the adenovirus. Between 1980 and 1985 our research focused on the analysis of the structure, expression and functions of early viral genes and gene products including those that enable the adenovirus to transform nonpermissive cells. Several areas have received special attention, including (I) the expression of viral tumor-antigens in adenovirus-transformed and -induced tumor cells, (II) properties and functions of the viral gene products that cause malignant transformation, and (III) interactions of viral gene functions with the regulation of the host cell cycle. Our approach required a combination of techniques including *in vitro* translation of selected viral messenger RNA, protein biochemistry, peptide synthesis, immunoprecipitation, recombinant DNA, mutagenesis, and molecular genetics. In addition to the inherent challenge of understanding how adenovirus gene functions act in cellular transformation, these viruses have provided exciting model systems and tools for exploring the molecular biology of eukaryotic cells in general.

---
\* Institut für Genetik der Universität zu Köln
  (at present: Institut für Molekularbiologie (Tumorforschung) Universität Essen)

*Helmut Esche*

## 6.3.2    Summaries

### 6.3.2.1    Structure and Expression of Early Viral Gene Products in Adenovirus Type 12-Infected Human (KB, HELA) and -Transformed Rodent (Hamster, Mouse, Rat) Cells

Adenovirus early proteins are defined as those virus-coded proteins that are expressed prior to and in the absence of viral DNA replication. From genetic data it was already known that some of these early viral proteins can regulate transcription of viral and cellular genes in both a positive and a negative way [1, 2, 3, 4]. Other early proteins are required for viral DNA replication [5]. It was already known that the gene products of early region 1 are involved in virus-induced malignant cell transformation [6, 7]. However, the mechanism by which the viral oncogenes transform the nonpermissive cell is poorly understood. Extensive progress was made at this time in recognizing early viral genes, but it also became clear that there is still some way to go before the catalogue is complete. Our analysis of early adenovirustype 12 gene products was based (a) on *in vitro* translation of virus-specific mRNA (selected by molecular hybridization to cloned DNA-fragments) prepared from infected and transformed cells, using a cell-free system of rabbit reticulocytes, and (b) on immunoprecipitations, using antibodies directed against synthetic peptides or viral protein domains expressed as viral:beta-galactosidase fusion proteins in *Escherichia coli*. We had identified 21 adenovirus type 12 early proteins, which were encoded by five defined transcription units. A more detailed analysis was then performed for gene products expressed from early regions 1a and 1b that have been shown by genetic studies to be necessary and sufficient for virus-induced transformation. Data we obtained concerning the kinetics of synthesis during infection, the sub-cellular localisation, protein stability, posttranslational modifications, and cellular proteins co-precipitated with the viral E1a and E1b tumor antigens are summarized in Table 6.3.1.

A comparison of viral proteins expressed in infected cells with those detected in adenovirus transformed cells has shown, that in addition to early regions 1a and 1b, which are essential for the induction and maintenance of the transformed state, the conventional early regions 2a, 3, and 4 are expressed in these cells, whenever they are present and not at least partially deleted during or after the process of viral genome integration. In none of the adenovirus-transformed cell lines investigated so far, have functional mRNA and proteins been detected as encoded by the region E2b (which codes for proteins necessary for viral DNA replication) and other viral regions (which are expressed exclusively late in productive infection, i.e. virus structural proteins).

**Table 6.3.1**

| Region | Coordinates (Map units) | Proteins (MW) | Time of synthesis After infection (h) Appearance at early times | Presence in late times | Half-life (h) | Cellular localisation[x] nucleus | cytoplasm | Phosphory-lated | Co-precipitated cellular (c) and viral (v) proteins |
|---|---|---|---|---|---|---|---|---|---|
| E1a | 1.3–4.5 | 45kDa 38kDa/36kDa 30kDa/29kDa 26kDa/24kDa 24kDa/22kDa | 3 to 5 | present in decreased amounts | 1 to 1.5 | $++^y$ | $-/(+)$ | yes (ser) | 200kDa (c) 110kDa (c) 68kDa (c) 32kDa (c) 14,5kDa (c) |
| | | 34kDa | present *exclusively* in Ad12-transformed cells | | ca. 1 | + | – | yes (ser, tyr) | |
| E1b | 4.5–11.2 | 58kDa | 8 | present | ca. 3 (ca. 20)$^z$ | + | + | yes (ser, thr) | 70kDa (c) 53kDa$^z$ (c) 21kDa (c) |
| | | 19kDa (17kDa) | 8 | present in increased amounts | ca. 10 | + | ++ | yes (?) | |

[x] Cytoplasm refers to cytosol plus membranes
[y] Present in large amounts
+ present in normal amounts
– undetectable amounts
[z] In adenovirus type 12-transformed cells

281

Three observations should be discussed here in more detail:

a) As it was known from S1-nuclease mapping of transcripts from this region that 6 only three, possibly four, differently spliced mRNAs are transcribed from early region 1a [8], we were surprised to discover so many proteins (more than 12 polypeptides) encoded by early region 1a). By translating E1a mRNA in cell-free systems under conditions where newly synthesized polypeptides cannot be phosphorylated and by performing peptide analyses of E1a proteins from infected cells, we and others have demonstrated that a single primary translation product is produced from each mRNA, and that the polypeptides are then specifically modified by phosphorylation and proteolytic events.

b) A comparison of E1a polypeptides immunoprecipitated from infected cells with different adenovirus type 12-transformed cell lines indicated an additional novel E1a protein of 34 kDa expressed in all *in vitro* transformed and *in vivo* induced tumor cells. We could never detect this E1a polypeptide after *in vitro* translation of pre-selected viral mRNA prepared from infected human or infected rodent cells. Mapping of the E1a mRNAs from Ad12-transformed cells using both S1- and mung bean nuclease assays gave first indications that the 34 kDa E1a polypeptide was encoded by an mRNA which was exclusively spliced in virus-transformed cells. We have currently cloned this mRNA in a recombinant retrovirus expression vector to investigate the function of this tumor-antigen.

c) Immunoprecipitation experiments using antibodies directed against E1 proteins indicated complex physical interactions of the viral proteins with cellular polypeptides. Several cellular proteins with molecular weights between 200 kDa and 14 kDa have been co-precipitated using anti-E1a antibodies. One of these proteins, a 105 kDa polypeptide, has now been identified by others as an anti-oncogene protein encoded by the Retinoblastoma (Rb) gene [9], whose homozygous inactivation triggers retinoblastomas and osteosarcomas [10]. This finding and the physical interaction of this anti-oncogene protein with an adenovirus tumor-antigen attracts interest because the Rb genes in question appears to act by restraining the growth of normal cells.

We have also been able to show that, in Ad12-transformed cells, the E1b 58 kDa tumor-antigen is physically associated with a cellular 53 kDa (p53) phosphoprotein. In virus-infected cells, however, the same viral protein is associated with a 21 kDa polypeptide encoded by early region 4 [11]. The cellular p53 protein has also been shown to be associated with the large E1b T-antigen of Ad2 and the large T-antigen of the SV40 [12]. The fact that the same cellular protein is associated with the large T-antigens of both tumor viruses in their respective transformed cells indicates that the two viral proteins may share some common functions. The cellular p53 protein occurs in increased amounts in a variety of *in vitro* transformed and *in vivo* induced tumor cells, compared to their non-transformed counterparts. The elevated concentration of the p53 protein may be significant to the transformed state of those cells. In non-transformed cells the half-life of this cellular protein is very short (20 min), while it is extremely long (about 20 h) when the protein is associated with the tumor antigens of SV40 or adenoviruses. The above observation has led to speculations about the nature and function of the physical association between the viral tumor antigens and the p53 protein, which seems to stabilize this cellular protein by reducing its rate of degradation or turnover. The cellular p53 protein also has an elevated half-life in tumor cells induced by other agents (mutagens, high energy radiation) than tumorvirus infection. There is accumulating evidence that the p53 protein is the second anti-oncogene known today [13]. This information might suggest that a physical association with a viral tumor antigen inhibits the repressor-like function of this cellular protein. This mechanism could play a relevant role in the regulation of cell proliferation of tumorvirus-infected cells.

## 6.3.2.2   Cell Factor(s) Mediated DNA-Binding of an Adenovirus Type 12 Coded Tumor-Antigen

Functions of the adenovirus E1a proteins were elucidated by employing point and deletion mutants in the E1a genes. Such virus mutants fail to transcribe all other early viral genes and are unable to induce or repress the expression of specific cellular genes after infection. The mechanisms mediating the positive and negative regulation of transcription by E1a proteins are unknown at present. Whether these proteins interact with the RNA polymerase II, transcription factors, or counteract a cellular repressor bound to viral or cellular DNA, remains to be seen.

The aim of this project was to determine if one of the viral E1a proteins could bind to double stranded (ds) DNA. Nuclear extracts were prepared from infected and transformed cells and were tested for binding of E1a proteins to viral and cellular dsDNA-cellulose. Each fraction of the salt gradients eluted from the DNA-cellulose columns was screened for E1a polypeptides by immunoprecipitation with E1a specific antibodies. All E1a proteins, except a 38/40 kDa polypeptide, were detected in the flow through. However, the latter one eluted from the column with 0.35 M KCl. In order to investigate the nature of the possible DNA binding of this E1a polypeptide, we repeated the experiments using partially fractionated nuclear protein extracts. Surprisingly, we could not find any DNA-binding of the partially purified 38/40 kDa protein. Incubation of the partially

purified protein fraction containing the E1a polypeptide with freshly prepared nuclear extracts from uninfected cells, however, restored the capabi-
lity of this E1a polypeptide to bind to dsDNA. These results suggested
that the binding of the E1a 38 kDa/40 kDa protein is mediated by one or
more cellular proteins. These observations have been confirmed by others
[14]. Filter binding assays using cloned viral and cellular DNA fragments
carrying promoter and enhancer sequences did not indicate any sequence-
specific DNA-binding of these viral/cellular protein complexes. The retinoblastoma 105 kDa polypeptide and the transcription factor AP1, which
both physically interact with E1a polypeptides and for which it has been
shown recently that they bind to dsDNA [15, 16], might be candidates for
participation in the observed DNA-bound protein complex. Our current
studies are focusing on the analysis of the cellular proteins in this pro-
tein/DNA complex.

## 6.3.2.3   Interaction of Adenovirus Gene Functions with the Regulation of the Host Cell Cycle

It has been known for some time that adenoviruses induce cellular DNA
synthesis in quiescent non-permissive or semi-permissive cells and cause
alterations in the cell cycle progression of growing cells [17]. The virus-
induced cell cycle appears to be abnormal; DNA replication is uncoupled
from the synthesis of ribosomal RNA [18], numerous chromosomal aberrations are induced in G2 phase, and abnormal chromosome segregations
are visible after mitosis [19]. Finally, adenoviruses of all serotypes can
immortalize non-permissive primary cell *in vitro*, which is an early and
essential step in cell transformation. Infection of primary cells with adenovirus mutants or transfection of specific viral DNA fragments into these
cells have emphasized the role of the adenovirus early regions 1a and 1b
functions in altering the host cell cycle [20, 21]. The mechanism by which
viral tumor-antigens immortalize primary cells is poorly understood. As a
first step toward the analysis of the functions of adenovirus tumor-antigens
in cell cycle regulation we isolated 17 temperature-sensitive (ts) cell cycle
mutants of human KB cells after mutagenesis and analyzed their growth
behavior after infection with adenoviruses. Two of these mutants (tsC13,
tsSC8), which we characterized in more detail, grew well at 33°C, but
showed greatly reduced DNA synthesis at the non-permissive temperature
(40°C) and arrest in phase G1 of the cell cycle. The mutations are located
in different complementation groups. Infection of both mutants with Ad2-
or Ad12-wildtype virus at the non-permissive temperature induces cellular
DNA synthesis in the G1 arrested cells leading to cell cycle progression.
Experiments with Ad5 deletion mutants (d1312, d1313) have indicated

that functions of both early E1 regions, E1a and E1b, are required to induce DNA synthesis in these mutants. A transformation-deficient Ad12 mutant (d1SC-1), however, which carries a deletion of 68 bp at the 3' end in the first exon of early region 1a, is able to induce cellular DNA synthesis at 40°C. These observations indicate that the functions of the E1a tumor-antigens responsible for immortalization and transformation can be separated.

Both mutants support replication of adenovirus types 2 and 12 at the permissive temperature (33°C) and replication of Ad2 at the non-permissive temperature. Ad12 DNA replication, however, is totally blocked in mutant tsSC8 at 40°C, while it remains substantially unaffected in mutant tsC13. This observation may indicate that for replication of the adenovirus type 12 genome an additional cellular factor is required, which is affected in some way in this mutant.

We are currently investigating which of the E1a and E1b gene products are responsible for releasing the block in phase G1 of the cell cycle in these mutants. Another goal of our laboratory work is to identify the cellular genes affected in these ts cell-cycle mutants and to understand their functional role in the regulation of the Go/G1- to S-phase transition in the cell cycle.

# 6.3.3   References

[1] A.J. Berk, F. Lee, T. Harrison, J. Milliams, P.A. Sharp: Pre-early adenovirus 5 gene product regulates synthesis of early viral messenger RNAs. Cell 17 (1979) 935–944.

[2] J.R. Nevins: Induction of the synthesis of a 70.000 Dalton mammalian heat shock protein by the adenovirus E1a gene product. Cell 19 (1982) 913–919.

[3] E. Borrelli, R. Hen, P. Chambon: Adenovirus-2 E1A products repress enhancer-induced stimulation of transcription. Nature 312 (1984) 608–612.

[4] A. Velcich, E.B. Ziff: Adenovirus E1a proteins repress transcription from the SV40 early promoter. Cell 40 (1985) 705–716.

[5] B.W. Stillman, F. Tamanoi: Adenovirus DNA replication: DNA sequences and enzymes required for initiation in vitro. Cold Spring Harbor Symp. Quant. Biol. 47 (1983) 741–752.

[6] A.J. van der Eb, H. van Ormondt, P.I. Schrier, J.H. Lupker, H. Jochemsen, P.J. van den Elsen, R. DeLeys, J. Mast, P.C. van Bereren, R. Dukema, A. DeWaard: Structure and function of the transforming genes of human adenoviruses and SV40. Cold Spring Harbor Symp. Quant. Biol. 44 (1979) 383–399.

[7] H.E. Ruley: Adenovirus early region 1A enables viral and cellular transforming genes to transform primary cells in culture. Nature 304 (1983) 602–606.

[8] Y. Sawada, K. Fujinaga: Mapping of adenovirus type 12 mRNAs transcribed from the transforming region. J. Virol. 36 (1980) 639–651.

[9] P. Whyte, K.J. Buchkovich, J.M. Horowitz, S.H. Friend, M. Raybuck, R.A. Weinberg, E. Harlow: Association between an oncogene and an anti-oncogene: The adenovirus E1A proteins bind to the retinoblastoma gene product. Nature 334 (1988) 124–129.

[10] S.H. Friend, R. Bernards, S. Rogelj, R.A. Weinberg, J.M. Rapaport, D.M. Albert, T.P. Dryja: A human DNA segment with properties of the gene that predisposes to retinoblastoma and osteosarcoma. Nature 323 (1986) 643–646.

[11] P. Sarnow, P. Hearionig, C. Anderson, D.H. Halbert, T. Shenk, A.J. Levine: Adenovirus E1b 58k tumor antigen is physically associated with an E4–25kd protein in adeno-virus productively infected cells. J. Virol. 36 (1984) 642–647.

[12] P. Sarnow, Y.S. Ho, J. Williams, A.J. Levine: Adenovirus E1b-58kd tumor antigen and SV40 large tumor antigen are physically associated with the same 53kd cellular protein in transformed cells. Cell 28 (1982) 387–394.

[13] A. Finlay, P.W. Hinds, A.J. Levine: The p53 protooncogene can act as a suppressor of transformation. Cell 57 (1989) 1083–1093.

[14] J.-L. Ko, B.L. Dalie, E. Goldman, M.L. Harter: Adenovirus-2 early region 1A protein synthesized in Escherichia coli extracts indirectly associates with DNA. EMBO J. 5 (1986) 1645–1651.

[15] W.-H. Lee, J.-Y. Shew, F.D. Hong, T.W. Sery, L.A. Donoso, J.-L. Young, R. Bookstein, E.Y.-H. Lee: The retinoblastoma susceptibility gene encodes a nuclear phosphoprotein associated with DNA binding activity. Nature 329 (1987) 642–645.

[16] W. Lee, P. Mitchell, R. Tjian: Purified transcription factor AP-1 interacts with TPA-inducible enhancer elements. Cell 49 (1987) 741–752.

[17] M. Rossini, R. Weinmann, R. Baserga: DNA synthesis in temperature sensitive mutants of the cell cycle infected by polyoma virus and adenovirus. Proc. Natl. Acad. Sci. USA 76 (1979) 4441–4445.

[18] A.W. Braithwaite, J.D. Murray, A.J.D. Bellett: Alterations to controls of cellular DNA synthesis by adenovirus infection. J. Virol. 39 (1981) 331–340.

[19] J.D. Murray, A.J.D. Bellett, A.W. Braithwaite, L.K. Waldron, I.W. Taylor: Altered cell cycle progression and aberrant mitosis in adenovirus-infected rodent cells. J. Cell. Physiol. 111 (1982) 98–96.

[20] A.W. Braithwaite, B.F. Cheetham, P. Li, R.C. Parish, L.K. Waldron-Stevens, A.J.D. Bellett: Adenovirus-induced alterations of the cell growth cycle: a requirement for expression of E1A but not of E1B. J. Virol. 45 (1983) 192–199.

[21] S. Stabel, P. Argos, L. Philipson: The release of growth arrest by microinjection of adenovirus E1A DNA. EMBO J. (1985) 2329–2336.

# 6.3.4 Publications

H. Esche, M.B. Mathews, J.B. Lewis: Proteins and messenger RNAs of the transforming region of wild type and mutant adenoviruses. J. Mol. Biol. 142 (1980) 393–417.

H. Esche: Viral gene products in adenovirus type 2 transformed hamster cells. J. Virol. 41 (1982) 1074–1082.

H. Esche, B. Siegmann: Expression of early viral gene products in adenovirus type 12-infected and -transformed cells. J. Gen. Virol. 60 (1982) 99–113.

A. Starzinski-Powitz, M. Schulz, H. Esche, N. Mukai, W. Doerfler: The adenovirus type 12-mouse cell system: Permissivity and analysis of integration pattern of viral DNA in tumor cells. EMBO J. 1 (1983) 493–499.

H. Esche, H. Lubbert, B. Siegmann, W. Doerfler: The translational map of the Autographa california nuclear polyhydrosis virus (AcNPV) genome. EMBO J. 12 (1983) 1629–1637.

H. Esche, M. Reuther, K. Schughart: Early and late proteins of adenovirus type 12: Translation mapping with RNA isolated from infected and transformed cells. Curr. Top. Microbiol. Immunol. 111 (1984) 91–106.

K. Schughart, E. Bause, H. Esche: Structure and expression of adenovirus type 12 E1b 58K protein in infected and transformed cells: Studies using antibodies directed against a synthetic peptide. Virus Res. 3 (1985) 41–56.

H. Esche: Early proteins coded by the adenovirus genome: Functional aspects. In: W. Doerfler (ed.): Adenovirus DNA: The Viral Genome and its Expression. Martinus Nijhoff Publishing, Boston-Dordrecht-Lancaster 1986, pp. 223–269.

## Survey of Doctoral Students

Marietta Brotz-Reuther (1979–1981): Adenovirus Type 12 late proteins: Identification and mapping.

Klaus Schughart (1983–1986): Analysis of the structure and the expression of the adenovirus type 12 E1b 58kDa protein using antibodies directed against fusion proteins and synthetic peptides. Postdoctoral Fellow, Yale University.

Monika Schrader (1983–1987): Temperature sensitive cell cycle mutants: Induction of cellular DNA-synthesis in G1-arrested cells by adenovirus infection. Postdoctoral Fellow, Universität Düsseldorf, FRG.

# 6.4 Replication of an Alphavirus in Cultured Vertebrate and Insect Cells

Dennis T. Brown*

## 6.4.1 Introduction

The alphaviruses are among the simplest of membrane-containing viruses. The most intensively studied members of this family are Semliki Forest virus (SFV) and the prototype, Sindbis virus. Since the production of temperature-sensitive (ts) mutants of Sindbis virus by Elmer Pfefferkorn and Boyce Burge in the mid 1960's, this virus has served as an important model membrane-containing animal virus. Considerable information has been gathered on the intracellular processing of virus structural proteins and on the events surrounding the envelopment of the assembled virus nucleocapsid in the host cell plasma membrane. The alphaviruses have also served as an important model system for investigations of the process by which membrane containing viruses penetrate cells (for a review, see Ref. [1]).

Sindbis is an excellent model system for investigations on the replication and assembly of a membrane-containing virus. It is simple in its composition, containing single-stranded plus polarity RNA, three virus coded proteins (two glycosylated membrane proteins and one basic core protein) and a membrane bilayer which is derived from the host cell plasma membrane. The viral structural proteins are produced from a subgenomic RNA (26S) as a single polyprotein (Fig. 6.4.1). The sequence of the components of this polyprotein is $NH_2$, C, $E_3$, $E_2$, $E_1$, COOH [2]. The capsid protein is cleaved from the developing polypeptide chain by a proteolytic activity residing in its own sequence. The capsid protein and the envelope proteins then undergo separate pathways of maturation. The capsid protein combines with 42S progeny RNA and condenses into an icosahedron. The

---

* Institut für Genetik der Universität zu Köln
  (at present: Cell Research Institute and the Department of Microbiology –
  The University of Texas at Austin)

rich cytoplasmic vesicles appearing in mosquito cells soon after infection by Sindbis virus. Virus nucleocapsids in the cytoplasm of the infected cells were rare. Instead, an accumulation of naked and membrane-bound nucleocapsids occured within these vesicles. Partially enveloped structures consisting of membrane fragments with attached nucleocapsids were found in the media of the infected cells. Thus, it appeared that in this pathway the viral envelope is obtained by the budding of nucleocapsids through virus protein modified intravasicular membranes, and the release of mature virions occurs upon the fusion of the vesicular membrane with the host plasma membrane in an exocytic process.

We also found that invertebrate cells differed from vertebrate cells in their response to Sindbis virus infection [18]. Upon infection, vertebrate cells will typically pass through an acute phase of infection, releasing large quantities of virus during which a decrease in host macromolecular synthesis occurs. The synthesis of alphavirus RNA is required for the inhibition of host protein synthesis and DNA synthesis. Although most vertebrate cultures do not survive infection with alphavirus, some reports of vertebrate cell populations persistently infected with alphaviruses exist in the literature. Several hypotheses have been developed to explain the cause of the cytopathic effect seen in the vertebrate cells. These involve various mechanisms of arresting host protein synthesis at the level of translation and by a shutdown of host messenger synthesis. It seems likely that a combination of events is responsible for the development of the cytopathic effect in the vertebrate cell.

In the mosquito cell, the development of a cytopathic response to infection by Sindbis varies, depending upon the species specific cell line [19, 20, 21]. The U4.4 cell line of *Aedes albopictus*, which was produced by cloning in my laboratory, produces large amounts of virus in the acute phase of infection without detectable cytopathic effect. In contrast, the LTC7 cell line, produced from the same parent cell population by Victor Stollar and employed in studies in my laboratory, also produces large quantities of virus and displays extensive cytopathic effects when infected with Sindbis virus. The events involved in the establishment of cytopathic effects in the invertebrate cell may be similar to those in the vertebrate, but this has not been demonstrated. It is possible that in the U4.4 cell line, protection from the cytotoxic effects of the virus is by the compartmentalization of the virus within the cytoplasmic vesicles described above.

Upon the infection of vertebrate and invertebrate cells with Sindbis virus, an acute phase of infection with peak virus production occurring by 12 to 14 h post-infection was reported by my laboratory in the early 1970's [17]. In the vertebrate system, cell death and lysis generally occur by 24 h post-infection. However, in the mosquito cell system, these high levels of virus production may continue for up to 5 days post-infection, at which time release of the virus begins to decline until only low levels of virus are produced and released into the media. B. Riedel, in my laboratory, examined the percentage of cells infected by alphaviruses (as determined by infectious center and immunofluorescence assays) in acutely and persistently infected cultures of mosquito cells and demonstrated a decline from 100% cells infected in the acute phase to less than 10% cells infected in the persistent phase [18]. The decline in virus production occurs concomitantly with the decline in infected cell number and a decrease in viral RNA synthesis in the shift from the acute to persistent phase. As this shift occurs, the frequency of detection of the virus-containing cytoplasmic vesicles by electron microscopy also decreases. In contrast to persistently infected vertebrate cultures, persistently infected invertebrate cultures do not undergo periodic crisis. B. Riedel found that when the rates of growth in the general cellular morphology of uninfected and persistently infected mosquito cell populations were compared to one another, no differences were observed.

Several hypotheses have been presented providing mechanisms for generating and maintaining persistent infections [3]. One of the most prevalent involves interference with the replication of wild-type virus by defective interfering particles with the net result that fewer cells will become productively infected. Although defective (truncated) viral RNA is synthesized in alphavirus infected mosquito cells, this RNA is not packaged or apparently released from the cells until after the shift to the persistent phase of infection has taken place. Furthermore, it has been demonstrated that defective interfering virus-produced invertebrate cells do not interfere with virus replication in insect cells. Thus, it would seem unlikely that defective virus particles are involved in the establishment of the persistent stage of infection. Temperature-sensitive mutants of virus, which may also play a role in the establishment of persistent infections in the invertebrate cultures, appear in infected mosquito cell cultures but are unlikely candidates for involvement in the establishment of persistence as they generally appear after the shift from the acute to the persistent phase. An additional argument against the role for either these mutants or defective particles in the establishment of persistence is suggested by the requirement for extracellular virus to maintain the infected state of the culture. B. Riedel demonstrated that persistently infected cultures may be cured of virus by incubation in virus-specific antisera, implying that cells within a culture must be reinfected to maintain the persistently infected state of the colony [18].

It has been demonstrated that invertebrate cell culture antiviral activity of interferons can participate in the establishment of persistent infections. However, despite many attempts, interferon has never been observed in invertebrate cells. The mode of action of antiviral agents (interferons) produced by infected vertebrate cells has been thoroughly examined. As indicated above, agents of this type have not been identified in infected invertebrate cells. B. Riedel and I, working in the laboratory in Köln, discovered a low molecular weight antiviral protein in the media of *Aedes albopictus* cells infected with Sindbis virus [19]. Treatment of mosquito cells with this antiviral agent prior to infection resulted in levels of virus production and percentages of infectious centers characteristic of persistently infected cell populations. It was found that the effectiveness of the agent in inhibiting Sindbis virus production was improved by incubating the mosquito cells in media containing the activity prior to infection. We found that this agent was insensitive to antisera raised to Sindbis virus produced in BHK cells and was rapidly inactivated by incubation at 56°C. The Sindbis virus-induced antiviral agent was found to be sensitive to proteinase K but insensitive to alpha and beta galactosidases, phospholipases, and a battery of DNAs and RNAs. We found that the antiviral activity was initially detectable in the media of Sindbis virus-infected mosquito cells at the end of the acute phase of infection (ca. 42 h postinfection). The agent is found at higher levels as the cells shift more and more into the persistent phase. Therefore, either the activity or the amount of the antiviral protein increases with the length of infection with Sindbis virus. The most striking feature of this antiviral protein was the extent of cell and virus specificity. The Sindbis virus-induced agent does not interfere with Sindbis production in BHK cells, and no inhibition with the replication of Semliki Forest virus or West Nile virus occurred in mosquito cells treated with the protein.

Since the discovery of these agents by B. Riedel and me in the late 1970's, Lynn Delgarno and his co-workers in Australia have identified an antiviral agent produced by Semliki Forest virus-infected mosquito cells [20]. This antiviral agent is also cell and virus specific and insensitive to serums specific for the virus, sensitive to proteinase K, and heat sensitive. Using actinomycin D, Delgarno and co-workers demonstrated that the Semliki Forest virus-induced antiviral protein requires host transcription and inhibits RNA in protein synthesis. Since the original report of the identification of the antiviral protein by B. Riedel and me in Köln, we have examined cell specificity of this agent with respect to a number of clones of *Aedes albopictus* cells. Several persistently infected cell lines derived from U4.4 cells were shown to release large amounts of this specific antiviral activity in the surrounding media. Persistently infected C710 cells (derived from the same parent cell culture), on the other hand, do not produce an antiviral activity similar to that observed with Sindbis virus infected U4.4 cells. Virus production in Sindbis virus-infected U4.4 cells and in C636 cells (another clone of *Aedes albopictus* cells) was inhibited by a 48 h preincubation in the antiviral agent, produced by persistently infected U4.4 cells [21]. However, Sindbis virus production in this cytopathic effect positive C710 cell line was uninfected by the same treatment. We have shown that the amount of viral RNA present in Sindbis virus infected U4.4 cell cultures treated with the antiviral agent released from the persistently infected cell population is much lower than that obtained in infected controls incubated in the medium of uninfected cells. Treatments of 72 and 48 h prior to infection resulted in the greatest reduction in RNA levels [21]. Two possible explanations were proposed for these observations. Either viral RNA synthesis is inhibited, or viral RNA is degraded in cells incubated in the antiviral activity. If viral RNA synthesis is inhibited by the antiviral agent, the level at which this inhibition occurs is unclear. Alterations in the efficiency with which the viral genome is uncoated, translation of nonstructural proteins, or production of the negative strand could all inhibit viral RNA replication. However, the stunning specificity with respect to the agent's ability to specifically suppress Sindbis virus replication while allowing for replication of the closely related Semliki Forest virus is totally obscure.

Indirect evidence suggests that the inhibitory action of the antiviral agent is specific for viral RNA synthesis [21]. Persistently infected cells, which produce the antiviral agent, continue to divide and grow at rates comparable to uninfected cells. No obvious changes occur in U4.4 cells incubated in the antiviral agent. If an alteration of cellular RNA synthesis is induced by the antiviral agent, it must either be of a minor nature or reversible. In addition, the unique virus specificity exhibited by these antiviral agents produced by infected mosquito cells indicates a specific interaction with the virus-replicating complex.

The conduct of these projects, which were initiated in Köln between 1973 and 1978, continues in Austin, Texas. As this document is being prepared, we are continuing research on intracellular transport of virus protein by isolating and characterizing pre- and post-Golgi transport vesicles. We have initiated a project to produce a three-dimensional reconstruction of the Sindbis virus nucleocapsid from images obtained by cryoelectron microscopy. We have also succeeded in producing crystals of Sindbis virus produced from insect cells. These crystals will be used for X-ray crystallographic studies of the virus structure at high resolution.

## 6.4.4   References

[1]   S. Schlesinger, M. J. Schlesinger: The Togaviruses. Academic Press, New York 1986.
[2]   E. G. Strauss, J. H. Strauss: Structure and replication of the alphavirus genome in the togaviridae and flaviviridae. Plenum Press, New York-London 1986, pp. 35–90.
[3]   D. T. Brown, L. Condreay: Replication of alphavirus in mosquito cells. In: S. Schlesinger (ed.): The Viruses. Academic Press, New York 1985, pp. 473–501.

[4] D. T. Brown, M. R. F. Waite, E. R. Pfefferkorn: Morphology and morphogenesis of Sindbis virus as seen with freezeetching techniques. J. Virol. 10 (1972) 534–536.

[5] D. Renz, D. T. Brown: Characteristics of Sindbis virus temperature-sensitive mutants in cultured BHK-21 and *Aedes albopictus* (mosquito) cells. J. Virol. 19 (1976) 775.

[6] D. T. Brown, J. F. Smith: Morphology of BHK-21 cells infected with Sindbis virus temperature-sensitive mutants in complementation groups D and E. J. Virol. 15 (1975) 1262.

[7] J. F. Smith, D. T. Brown: Envelopment of Sindbis virus: synthesis and organization of proteins in cells infected with wild-type and maturation defective mutants. J. Virol. 22 (1977) 662.

[8] D. T. Brown: The assembly of alphaviruses. In: R. W. Schlesinger (ed.): The Togaviruses: Biology, Structure, Replication. Academic Press, New York 1980, pp. 473–501.

[9] C. Erwin, D. T. Brown: Intracellular distribution of Sindbis virus membrane proteins in BHK-21 cells infected with wild-type virus and maturation-defective mutants. J. Virol. (1980) 775–786.

[10] H. Scheefers, U. Scheefers-Borchel, J. Edward, D. T. Brown: Distribution of virus structural proteins and protein-protein interactions in plasma membrane of baby hamster kidney cells infected with Sindbis or vesicular stomatitis virus. Proc. Natl. Acad. Sci. USA (1980) 7277–7281.

[11] M. Knipfer, D. T. Brown: Intracellular transport and processing of Sindbis virus glycoproteins. Virol. 170 (1987) 117–122.

[12] J. F. Presley, D. T. Brown: The proteolytic cleavage of $PE_2$ to envelope glycoprotein $E_2$ is not strictly required for the maturation of Sindbis virus. J. Virol. 63 (1989) 1975–1980.

[13] K. Coombs, B. Brown, D. T. Brown: Evidence for a change in capsid morphology during Sindbis virus envelopment. Virus Res. 1 (1984) 297–302.

[14] K. Coombs, D. T. Brown: Topological organization of Sindbis virus capsid protein in isolated nucleocapsids. Virus Res. 7 (1987) 131–149.

[15] K. Coombs, D. T. Brown: Form-determining functions in Sindbis virus nucleocapsids: nucleosomelike organization of the nucleocapsid. J. Virol. 63 (1989) 883–891.

[16] K. Coombs, D. T. Brown: Organization of the Sindbis virus nucleocapsid as revealed by bifunctional cross-linking agents. J. Mol. Biol. 195 (1987) 359–371.

[17] J. B. Gliedman, J. F. Smith, D. T. Brown: Morphogenesis of Sindbis virus in cultured *Aedes albopictus* cells. J. Virol. 16 (1975) 913–926.

[18] B. Riedel, D. T. Brown: The role of extracellular virus in the maintenance of the persistent infection induced in *Aedes albopictus* (mosquito) cells by Sindbis virus. J. Virol. 23 (1977) 554–561.

[19] B. Riedel, D. T. Brown: A novel antiviral activity found in the media of Sindbis virus persistently infected mosquito (*Aedes albopictus*) cell cultures. J. Virol. 29 (1979) 51–60.

[20] S. E. Newton, L. Delgarno: Antiviral activity released from *Aedes albopictus* cells persistently infected with Semliki Forest virus. J. Virol. 47 (1983) 652–655.

[21] L. Condreay, D. T. Brown: Suppression of RNA synthesis by a specific antiviral agent produced in Sindbis virus infected *Aedes albopictus* (mosquito) cells. J. Virol. 62 (1987) 346–348.

# 6.4.5   Publications

The following publications were produced from research conducted in my laboratory, or through collaborations at the Institut für Genetik from July 1973 to January 1978.

D. T. Brown, J. B. Gliedman: Morphological variants of Sindbis virus obtained from infected mosquito tissue culture cells. J. Virol. 12 (1973) 1535–1539.

W. Doerfler, H. Burger, J. Ortin, E. Fanning, D. T. Brown, M. Westphal, U. Winterhoff, B. Weiser, J. Schick: Integration of adenovirus DNA into the host cell genome. Cold Spring Harbor Symp. Quant. Biol. 34 (1974) 505.

B. T. Burlingham, D. T. Brown, W. Doerfler: Incomplete particles of adenovirus types 2 and 12. I. Characteristics of the DNA associated with the incomplete adenovirions. Virol. 60 (1974) 419–430.

J. Groneberg, D. T. Brown, W. Doerfler: Uptake and fate of the DNA of adenovirus type 2 in KB cells. Virol. 64 (1975) 115.

D. T. Brown, J. F. Smith: Morphology of BHK-21 cells infected with Sindbis virus temperature-sensitive mutants in complementation groups D and E. J. Virol. 15 (1975) 1262.

J. B. Gliedman, J. F. Smith, D. T. Brown: Morphogenesis of Sindbis virus in cultured *Aedes albopictus* cells. J. Virol. 16 (1975) 913–926.

D. T. Brown, J. F. Smith, J. B. Gliedman, B. Riedel, D. Filtzer, D. Renz: Morphogenesis of Sindbis virus in cultured mosquito cells. In: E. Kurstak, K. Maramorosch (eds.): Invertebrate tissue culture. Academic Press, New York 1975, pp. 35–48.

R. E. Johnston, D. R. Tovell, D. T. Brown, P. Faulkner: Interfering passages of Sindbis virus: concomitant appearances of interference, morphological variants and truncated viral RNA. J. Virol. 16 (1975) 951–958.

D. T. Brown, M. Westphal, B. T. Burlingham, U. Winterhoff, W. Doerfler: The structure and composition of the adenovirus type 2 core. J. Virol. 16 (1975) 366–387.

L. W. Black, D. T. Brown: Morphological observations of *E. coli* cells infected with T4 phage head and internal protein defective mutants. J. Virol. 17 (1976) 894–905.

B. Kemper, D. T. Brown: The function of gene 49 of bacteriophage T4. II. Intracellular development and structural analysis of very fast sedimenting DNA. J. Virol. 18 (1976) 1000–1015.

D. Renz, D. T. Brown: Characteristics of Sindbis virus temperature-sensitive mutants in cultured BHK-21 and *Aedes albopictus* (mosquito) cells. J. Virol. 19 (1976) 775.

J. Kania, D. T. Brown: The functional repressor parts of a tetrameric *Lac* repressor-B-galactosidase chimaera are organized as dimers. Proc. Natl. Acad. Sci. USA 73/10 (1976) 3529–3533.

D. T. Brown, R. Riedel: Morphogenesis of vesicular stomatitis virus: electron microscope observations with freezefracture techniques. J. Virol. 21 (1977) 601.

J. F. Smith, D. T. Brown: Envelopment of Sindbis virus: synthesis and organization of proteins in cells infected with wild-type and maturation defective mutants. J. Virol. 22 (1977) 662.

B. Riedel, D. T. Brown: The role of extracellular virus in the maintenance of the persistent infection induced in *Aedes albopictus* (mosquito) cells by Sindbis virus. J. Virol. 23 (1977) 554–561.

B. Riedel, D. T. Brown: A novel antiviral activity found in the media of Sindbis virus persistently infected mosquito (*Aedes albopictus*) cell cultures. J. Virol. 29 (1979) 51–60.

R. Dunker, D. T. Brown: A tubular subviral structure produced in adenovirus infected KB cells. J. Virol. 31 (1979) 568–574.

H. J. Eggers, B. Bode, D. T. Brown: Cytoplasmic localization of the uncoating of picornaviruses. J. Virol. 92 (1979) 211–218.

# 7    Gene Transfer

# Chromosome Mediated Transfer of Human Genes

Klaus Willecke*

From 1971 to 1973, we studied transport systems for citrate [1] and malate [2] in *Bacillus subtilis*. Genetic characterization of these bacteria was eased by the relatively well-characterized system of DNA mediated transformation. When Dr. Eric Eisenstadt entered my laboratory, we began to investigate the transport of DNA into competent *Bacillus subtilis* cultures. We found that competent cultures synthesize a protein that binds to single-stranded DNA, suggesting that it may play a role in uptake of DNA and/or integration into the genome [3]. In those years, cloned genes were not yet available. Since I wanted to extend our investigation to the DNA uptake into mammalian cells, I suggested this project to Dr. Frank Ruddle at Yale University in whose laboratory I worked from August 1973 to May 1974. Just at the beginning of my stay at Yale, however, McBride and Ozer [4] reported that the Chinese hamster gene coding for hypoxanthine phosphoribosyl transferase (HPRT) could be transferred via isolated metaphase chromosomes into mouse cells. Thus, I decided to use the system of isozyme gene markers that had been mapped to human chromosomes in Ruddle's laboratory for characterization of transferred chromosomal fragments. First, we estimated that less than 1% of the human genetic material or less than 20% of the human X-chromosome was transferred and maintained in the recipient genome when we selected for transfer of human HPRT into mouse cells [5]. When I came back from Yale to Köln, I studied in our new cell culture laboratory the chromosome mediated gene transfer of human cytosyl thymidine kinase (TK). In this case, we showed that two linked genes (thymidine kinase and galactokinase) could be transferred together into mouse recipient cells [6]. The distance between the two genes was estimated at that time to be shorter than 0.2% of the diploid human genome. We also discovered that phenotypically unstable gene transfer clones became stable during the selection process,

* Institut für Genetik der Universität zu Köln
  (at present: Institut für Genetik, Abteilung Molekulargenetik der Universität Bonn)

suggesting that the transferred genes became integrated into the recipient genome [6]. As a by-product of our studies of thymidine kinase activity in human cells, we assigned the human gene for mitochondrial thymidine kinase to human chromosomes 16 [7]. In the following years, we studied the integration (at the chromosome level) of human transgenomes into mouse cells after chromosome-mediated gene transfer. In all cases investigated, we found that the transferred chromosomal fragment segregated after cell hybridization with nonhomologous chromosomes. There was no evidence for homologous recombination as a mechanism of integration into the recipient genome [8, 9]. In 1979, we reported that an electrophoretic variant of mouse HPRT could be used for intraspecies *DNA mediated gene transfer* into mouse cells [10]. Shortly before, Wigler and associates had published their results on DNA mediated transfer of the mouse TK gene into TK deficient mouse cells [11]. These investigations formed the basis for DNA transfections into mammalian cells that subsequently became a standard method of mammalian molecular genetics when cloned genes and chromosomal fragments were characterized.

In retrospect it is interesting to compare a review on chromosomal gene transfer that I wrote in 1978 [12] with a more recent article from 1987 on the same subject [13]. Clearly our studies between 1973 and 1978 helped to work out the principles of gene transfer between mammalian cells. It is now well accepted that integration at non-homologous sites in the recipient genome is the preferred fate of genes transferred into mammalian cells. Only recently, the problem of selection for homologous recombination in the mammalian genome was tackled. In addition to DNA mediated transfections, chromosomal genes transfer still has great potential as a functional assay for complex genetic loci and for long-range human gene mapping [13].

# References

[1] P. Oehr, K. Willecke: Citrate-Mg2+ transport in Bacillus subtilis studied with 2-fluoro-L-erythro-citrate as a substrate. J. Biol. Chem. 249 (1974) 2037–2041.

[2] K. Willecke, R. Lange: C4-Dicarboxylate transport in Bacillus subtilis studied with 3-fluoro-erythro-malate as a substrate. J. Bacteriol. 117 (1974) 373–378.

[3] E. Eisenstadt, R. Lange, K. Willecke: Competent Bacillus subtilis cultures synthesize a denatured DNA binding activity. Proc. Natl. Acad. Sci. USA 72 (1975) 323–327.

[4] W. O. McBride, H. L. Ozer: Transfer of genetic information by purified metaphase chromosomes. Proc. Natl. Acad. Sci. USA 70 (1973) 1258–1262.

[5] K. Willecke, F. H. Ruddle: Transfer of the human gene for hypoxanthine-guanine phosphoribosyltransferase via isolated human metaphase chromosomes into mouse L-cells. Proc. Natl. Acad. Sci. USA 72 (1975) 1792–1796.

[6] K. Willecke, P. J. Davies, T. Reber: Cotransfer of two linked human genes into cultured mouse cells. Proc. Natl. Acad. Sci. USA 73 (1976) 1274–1278.
[7] K. Willecke, T. Reber, R. S. Kucherlapati, F. Ruddle: Human mitochondrial thymidine kinase is coded for by a gene on chromosome 16 of the nucleus. Somat. Cell Gen. 2 (1977) 237–245.
[8] P. Davies, K. Willecke: Segregation of human hypoxanthine phosphoribosyltransferase activity from somatic cell hybrids isolated after fusion of mouse gene transfer cells with Chinese hamster cells. Mol. Gen. Genet. 154 (1977) 191–197.
[9] K. Willecke, R. Mierau, A. Krüger, R. Lange: Chromosomal gene transfer of human cytosol thymidine kinase into mouse cells. Integration or association of the transferred gene with a nonhomologous mouse chromosome. Mol. Gen. Genet. 161 (1978) 49–57.
[10] K. Willecke, M. Klomfaß, R. Mierau, J. Doehmer: Intraspecies transfer via total cellular DNA of the gene for hypoxanthine phosphoribosyl transferase into cultured mouse cells. Mol. Gen. Genet. 170 (1979) 179–185.
[11] M. Wigler, A. Pellicer, S. Silverstein, R. Axel: Biochemical transfer of single-copy eucaryotic genes using total cellular DNA as donor. Cell 14 (1978) 725–731.
[12] K. Willecke: Results and prospects of chromosomal gene transfer between cultured mammalian cells. Theoretical and Applied Genetics 52 (1978) 97–104.
[13] D. J. Porteus: Chromosome mediated gene transfer: a functional assay for complex loci and an aid to human genome mapping. Trends in Genetics 3 (1987) 177–182.

## Students and Doctoral Co-workers between 1971 and 1978

I would like to thank all my co-workers – including those whose names are not mentioned below – for their enthusiasm and devoted collaboration in our laboratory in Köln.

Dr. Peter Oehr, Privatdozent at the Nuklearmedizinische Klinik der Universität Bonn.

Dr. Eric Eisenstadt, postdoc from the US, now associate professor at Harvard University, Boston.

Dr. Peter Davies, post-doctoral fellow from Great Britain, studied medicine afterwards and is now working as a physician.

Thomas Reber and Rudolf Mierau who worked during their diploma theses in my laboratory went on to get a M. D. and a Ph. D., respectively.

Dr. Wolfgang Wille, now associate professor at the Institut für Genetik der Universität zu Köln.